Charles Darwin

A Life from Beginning to End

(The Story of Darwin's Life How His Ideas Changed Everything)

Clyde Lee

Published By **Andrew Zen**

Clyde Lee

All Rights Reserved

Charles Darwin: A Life from Beginning to End (The Story of Darwin's Life How His Ideas Changed Everything)

ISBN 978-1-77485-734-2

No part of this guidebook shall be reproduced in any form without permission in writing from the publisher except in the case of brief quotations embodied in critical articles or reviews.

Legal & Disclaimer

The information contained in this ebook is not designed to replace or take the place of any form of medicine or professional medical advice. The information in this ebook has been provided for educational & entertainment purposes only.

The information contained in this book has been compiled from sources deemed reliable, and it is accurate to the best of the Author's knowledge; however, the Author cannot guarantee its accuracy and validity and cannot be held liable for any errors or omissions. Changes are periodically made to this book. You must consult your doctor or get professional medical advice before using any of the suggested remedies, techniques, or information in this book.

Upon using the information contained in this book, you agree to hold harmless the Author from and against any damages, costs, and expenses, including any legal fees potentially resulting from the application of any of the information provided by this guide. This disclaimer applies to any damages or injury caused by the use and application, whether directly or indirectly, of any advice or information presented, whether for breach of contract, tort, negligence, personal injury, criminal intent, or under any other cause of action.

You agree to accept all risks of using the information presented inside this book. You need to consult a professional medical practitioner in order to ensure you are both able and healthy enough to participate in this program.

Table Of Contents

Introduction _____ 1

Chapter 1: Development Of An
Unconventional Mind_____ 8

Chapter 2: The Voyage _____ 33

Chapter 3: Love And Prestige _____ 67

Chapter 4: Cambridge 1828-1831. _____ 94

Chapter 5: Residence Down From The 14th
Of Sept. 1842, Until The Present Time
1876. _____ 123

Chapter 6: I Just Loved Shooting! But I
Believe I Have To _____ 148

Introduction

"There is a greatness in this perspective of the world, and its many powers, after having been first breathed into a handful of shapes or even into one and that, as this planet has been moving around according to gravity's fixed law since its simple starting point, infinite forms of the most stunning and beautiful have evolved developed and are still being developed." Charles Darwin. Charles Darwin

"We should, however, admit, as it appears for me to be the case that a man with all the high-quality quality... nevertheless retains in his body the mark of his humble origins." Charles Darwin. Charles Darwin

Since the human mind was able to think it has been contemplating not just the significance of existence, but also the origins of the universe as well as the universe and all the natural wonders and beautiful forms of life within it. Even today the complexities of these issues remain the subject of contention and are typically summarized in the debate over the two theories of evolution and creationism.

On the other hand, there are those who insist the fact that God or some other divine being who designed and made every part of the universe as demonstrated by the multitude of myths about creation from different beliefs and traditions. One of the most popular myths of etiology is the classic tale about God, the Christian God who created the universe in just six days, and created man and woman from man's ribs and clay according to. This is a notion that people still believe in even today. Nearly 38 percent of Americans believed in the creation myth at the time of 2017, revealed by an Gallup poll. There are also less well-known stories that are not as well-known, like that of the Japanese story of creation, which explains how the gods and goddesses Izanagi and Izanami who were created from "elements combined with a single germ of life," shaped the Japanese islands using dirt and the sacred staff. The Mayans spoke of Tepeu, the creator of all things. They also praised Gucumatz as the "feathered spirit" who made the world by merely thinking and created on Earth the first group of human beings made of yellow and white corn.

However, there are those who adhere to the notion of as they say, "accept" evolution and scientific processes as facts. While captivating as these creation myths might be for those who oppose creationism who are agitating for their demise, there's no sound reason behind these old myths. The renowned theorist Stephen Hawking, arguably one of the most brilliant minds that ever lived, wrote, "Before we understand science it was natural to think the idea that God made the universe. However, science has an explanation that is more convincing." These words are shared by the famous evolutionary biologist, atheist and creator of the term "meme," Richard Dawkins in his book The Selfish Gene: "Today the evolution theory is just as susceptible to question as the idea that the earth revolves around the sun. ..."

"Some believe that evolution is merely the result of a hypothesis and that it is just an opinion," highly regarded scientist Neil deGrasse Tyson, another major proponent of evolution states. "The hypothesis of evolutionary theory - just like gravity theories is a scientific fact. Evolution did happen. Being able to recognize our kinship with the entire lifeform on Earth isn't just an actual science. According to me it's also an uplifting religious experience."

Some, however are in the middle. Maria Mitchell, America's first female astronomer was one of the characters mentioned above: "Scientific investigations, pushed forward and further will reveal new ways God is at work, and give us more profound revelations about the undiscovered." The 17th-century Polish scientist and astronomer Johannes Kepler, famed for his research into the movement of planets and his belief that humans could comprehend God's complicated and logical plans of the universe "The primary goal of any investigation into the outer world should be to uncover an orderly and harmonious system that was imposed on the world by God and that He has gave us through mathematical language." Sir Isaac Newton teetered between science and the concept of a God. "What we can see is a drop; what we do not understand can be an immense ocean" the scientist once confessed. "The amazing order as well as the harmony in our universe can only have been the design of an omniscient and all-powerful being."

In light of the positive discussions that take place in panels that are organized and the daily conversations to the often heated debates on message boards online or other online platforms it's possible to conclude these flames from this endless discussion are more intense and flaming than it has ever been. This year, an heated public debate on whether or not the concept of creation is an appropriate model for the origins of life was conducted between the science communicator and TV celebrity Bill Nye, on the side of evolutionaryists and Ken Ham, a Christian fundamentalist and young Earth creationist. All 900 seats inside the theater at the Cincinnati's Creation Museum, where the debate took place at the time, were sold out in the span of 48 hours. The eyes of over 3 million viewers that the two-hour debate attracted were riveted on their laptops, televisions and cell phones looking for their appointed spokespersons to defend their cause.

The main protagonist of this fervent existence-based debate Of course, the person who ignites this debate is the legendary Charles Darwin, who himself invented the term "creationists" and helped propel the controversial evolutionary theory to the limelight. Since he was adamantly against the grain in his research and was clearly an extremely divisive people of his time and would not hesitate to challenge one of the most fundamental beliefs of Western civilization. He wrote: "I cannot persuade myself that a benevolent and omnipotent God could have created parasitic wasps with the explicit purpose of feeding them within the living organisms of Caterpillars."

Although the majority of us today revere and admire Darwin, Darwin is still resented by those who believe he has poisoned many minds by promoting ludicrous and absurd beliefs that challenge what they consider to be the truthful declaration that is the word of God. Lord. With this in mind even though it was Darwin who ignited the flame but he's not the deity-obsessed, hardcore atheist who was obsessed with disproving God the way that many believe Darwin to be. What lies inside this intriguing man is a complex brainy, neurotic and suffering person, which is not surprising, makes the pioneer all more fascinating.

Chapter 1: Development of an Unconventional Mind

On February 12th in 1809 Mount House - a three-story red-brick Georgian mansion sat on the knoll that overlooks the Shropshire town of Shrewsbury as well as the River Severn - was vivified with the introduction of the addition of a family member. The man who built and was the master of the Mount House, a locally famous medical doctor by"Dr. Robert Waring Darwin, beamed with a beaming smile that spanned across his ear from one to the other, at the sight of his 5th son and second son. He was able to kiss his beloved wife, Susannah Hedgwood , on the forehead and thanked her for the cherished present. While the parents of their love gushed over their sleepy baby They decided to name him in honor of his uncle, who was a gifted medical student who was afflicted to a sudden, illness-related death. He was christened "Charles Robert Darwin."

The photograph of Jeremy Bolwell's Mount House Mount House

Robert Darwin

A lot was expected of infant Darwin Many within the community discussing the future of his fortunes and cerebral adventures before he even learned the art of speaking. His father, for instance had a successful self-made and was given nothing more than L20 from his great-grandfather, Erasmus, to establish his own business in Shrewsbury. Robert later tied the knot with an affluent lady and then further boosted his wealth by investing in property and paying hefty interest on loans that were extended to the wealthy. Darwin's parents were equally wealthy or better repute. The mother of Darwin was the pottery master and the inventor of the pyrometer as well as the "improved green glaze" Josiah Wedgwood, whose frequent patrons included the Queen Charlotte who was from England as well as Russian Empress Catherine of Russia. Great.

Erasmus Darwin

Erasmus was regarded as "one of the top intellectuals of the 18th the 18th century England," was an famous scientist, physician as well as author who was both celebrated and denigrated for his unconventional thoughts and "freethinking" methods. The clever and sharp mind that wrote Zoonomia ("The The Laws of Life") which delved into psychopathology, psychology and the anatomy that the body has was the same mind who wrote The Botanic Garden in two parts, which depicts an erotic depiction of the process of reproduction in plants. In a way, this work suggests the idea of evolution in the biological world, a idea that was a primitive one that would be further developed by his son and admirers. His "scandalous" views were inextricably linked with his unquestioned behavior like his denial of slavery, support for women's education and his choice to enjoy the company of homosexual friends, and his open attitudes towards sexual relations and even recommending it as a treatment for hypochondria. Even though his contemporary, the poet and philosopher Samuel Taylor Coleridge, hailed his work as "the most innovative of men who are philosophical ...[one who is in a fresh way about all subjects," his ideas were deemed as sexist to those in the Victorian society that they deliberately left out of the history books for more than 200 years.

Darwin as well as his four siblings Erasmus, Marianne, Caroline and Susan had the luxury of not ever having to endure one drop of poverty. 44-year old Robert had already established himself and could easily take care of his growing family. This Victorian period was plagued by scarlet fever and other infectious diseases, which made the sterilization process and the management of the estate, household and finances to be a full-time task which was handed over to the mother of the household.

Darwin was blessed by a large number of brothers and sisters and, although he was able to get to play with them in a jiffy the people around him identified Darwin as a quiet, calm child who was amused through his personal company. Darwin was a fan of miniature adventures on his quiet walks, looking at the animals and plants that he encountered, as well as collecting feathers, leaves nest pieces and other objects to keep for his personal collection as he went along. However, as sharp and sharp-witted as Darwin was, his intense fascination with animals occasionally led him into gruesome and dark depths, causing many critics to label him an alcoholic child. In addition to his own inclination of fabricating "deliberate falsehoods...done to the purpose of creating excitement" Darwin was often involved in thefts of a small amount. The most disturbing of them all was a case where the young boy "beat an innocent puppy...simply out of a sense strength." The boy later expressed regret for the incident.

Darwin at 7 years old.

There is speculation there was a devastating suicide of Susannah in 1817 which led to Darwin's troublesome behavior. After Susannah's death Robert was initially depressed. Robert focused on tasks, and left his oldest daughter, Marianne as well as Caroline, to manage the household matters. There was a change within Mount House - it might have been equally neat and luxurious however, not even the most powerful lamps can light the dim and dull house. Although it was home to seven people including Darwin's brand new niece, Emily Catherine, the home was cold and cold, and was made more cold due to Robert's temper issues and the depression of the children. Darwin's memory of his mother's death continued diminishing, and not just due to age, but also due to his father's grief and sisters refusing to speak about his mother's name.

In the end, the 17-year-old Caroline was Darwin's primary teacher throughout the beginning of his life. She taught her curious brother how to write and read and helped him master the basics of learning until the age of 7 when he was enrolled in the nearby day school. The school he attended for a year prior to transferring into Shrewsbury Grammar School, a institution of boarding that was overseen by the Reverend Samuel Butler, about a mile of Mount House. The school gave him an excellent education, with a focus in classics, including the analysis and grammar of Greek, Roman, and Latin texts. However, Darwin even though he developed an interest in Byron and Shakespeare was generally focused on his lack of interest in these subjects. In his autobiography Darwin declared that Darwin had been "considered by all his instructors and by his father as an ordinary young man, but not as smart as the average intelligence."

A photograph of an Darwin statue is unveiled in front of Shrewsbury Grammar School

But the teacher looked beyond his low test scores and saw the faintest hint of hope inside Darwin. He helped Darwin to nurture his passion for the natural world, and to build on his collection of beetles. While a child, Darwin also scavenged for bird eggs animals' shells, minerals as well as postage Franks. Darwin is known for shadowing the gardener in the family after classes, demonstrated an obvious passion for gardening. He also developed an unbeatable bond with his brother Erasmus, Jr., with the young scientists even turning a shed next to their home into a chemistry lab. The two of them took samples, conducted experiments and read numerous books - some of them that Darwin's professor recommended about botany, natural science, and historical expeditions, such as the one from Lewis and Clark.

Darwin continued to be prone to playing pranks and lying throughout the early years of high school. He irritated his classmates by pretending to find the apples hidden in strange places that Darwin himself had hidden earlier as well as once tricked a fellow student into believing that he could change the color of primroses using the help of a particular concoction of his creation. It was a good thing that Darwin eventually grew tired of his ego-driven machinations when the interest he craved was beginning to diminish.

Robert was not too enthusiastic about the mischief that his son was attempting to create regularly but he brushed it off as being nothing more than a game for kids however Darwin's mediocre academic results were beginning to cause him to worry. In addition to reworking his collection of peculiar objects Darwin was also a Darwin, a young Darwin was found wandering around the wooded areas of Shrewsbury using either the typical binoculars or shotguns in his hands and chasing birds in all shapes and sizes. "I am not convinced that any person could have displayed more devotion to the most sacred causes than the one I showed in hunting birds" Darwin later wrote. "How did I feel when I killed my first snipe...my exhilaration was so intense that I had a great deal of trouble getting my gun back in my hands shaking. The taste lasted for a long time and I soon became a proficient shooter."

While he was attempting to discredit the extraordinary love of his son for the natural world, Robert began to fear that Darwin was not able to pass on the traits of intelligence that were passed on to the rest of his family. He was worried Darwin would be nothing less than a mediocre hunter. While Darwin's enthusiasm for hunting only seemed to grow more intense with the passing of time. Not content with the variety of birds available that were available in Shrewsbury, Darwin and Erasmus began traveling to Wales and, at times they would travel to their cousins' home in Maer in order to participate in hunting sessions. His obsession with hunting is evident in the meticulously detailed ledger of sports that he meticulously updated. It included the number of birds and species of fowl and animals he shot down during every session.

It was in 1825 that Robert was so enraged by Darwin's antics, Robert ejected the boy of 16 from his school and attempted to re-instil some sense in Darwin. "You are only interested in shooting rats, dogs, and shooting," a disappointed Robert told Darwin. "You are a disgrace to yourself and your family members." The time was the right the time Robert said, to get Darwin to let go of his naiveté and think about his future.

In the hope of bringing the lazy teenager back into shape, Robert enrolled Darwin in his school of choice and The University of Edinburgh, and convinced him to study for the medical field. To encourage Charles to be on the lookout for his son's lackadaisical younger brother, Erasmus who was seeking a medical education from The University of Cambridge, Robert also sent Erasmus to Edinburgh in addition.

A modern representation that of Edinburgh's University of Edinburgh

The first time, Darwin and Erasmus, who resided in a cosy elegantly furnished boardhouse in 11, Lothian Street, enjoyed the new surroundings for a while. They took in a few lectures and joined an organization called the Royal Medical Society, but in the majority of cases they benefited from the school's vast and well-stocked library, acquiring and reading more than any of the other students. When Erasmus was ordered back to Shrewsbury after four months Charles, however Charles who found himself faced with the rocky water of the medical college by himself started to ignore his studies, choosing instead to read an unread book in the library every single day.

Darwin was fascinated by the intense satisfaction he felt by helping his father in the care of the elderly prior summer, however, Darwin was not impressed by the caliber of lectures at Edinburgh even on zoological topics which he dismissed as "intolerably boring." But be however Darwin was in his stay in Edinburgh that he began to dip his feet into the refreshingly energizing scientific pool, and as usual, did this outside his lecture room.

To begin, he bonded with and was a trainee to the renowned taxidermist John Edmonstone, a freed South American slave employed in the upholstery workshop at Edinburgh Museum. Edinburgh Museum. Instead of undergoing medical, surgical, or medical training, he sewed the tanned hides of beasts over wooden mounts, sewn together bird pelts and filled the animals with cotton balls, rags sawdust, sand and other substances similar to. He also learned to create eyes as well as other "soft tissue components" made of colored glass and painted wood. Prior to the introduction of taxidermytechniques, researchers relied on beautiful, but inaccurate drawings of animals in order to preserve the animals; but not just did the advent of the technique allow scientists to move just one step closer making precise 3-dimensional representations of the wonderful creatures in the world but it also gave them a technique that could prove useful for Darwin's future endeavours.

Alongside his taxidermy lessons, John Edmonstone captivated Darwin by his captivating accounts of the stunning Guyana Rainforest and its vibrant fauna and flora, which is home to 6,409 species plants, and 1,263 breeds of mammals, birds amphibians, reptiles, and mammals. Edmonstone studied Darwin's notebooks at his whim as well as taught Darwin's young mind how to identify the species and classes of colonial wildlife and the rock strata. Darwin expressed the satisfaction of his one-on-1 sessions with the taxidermist who was cultured in a letter to his sister Susan who was able to reveal a further advantage to his meetings together with Edmonstone: "I am going to learn how to stuff birds from the perspective of a blackamoor...it is a good idea to recommend inexpensiveness, if more than that, given that the price is only one dollar (about 1 L or $1.39 USD in the present) for an hour each day for two months."

Darwin was also a member and was ultimately granted entry to Darwin was also admitted to the Plinian Society, a prestigious natural history club located on the campus of the University established by professor Robert Jameson 2 years prior to the transfer. In a bid to explore his interests with like-minded pupils, Darwin took to the podium in March 1827 to present to his fellow members his observations that he had gathered on his travels with naturalist 34-year-old Robert Edmond Grant, a card-carrying fellow member with the Royal Society of Edinburgh (Scotland's national academy of sciences) as well as an "stalwart" in members of the Plinian Society.

Grant

Together, they scoured around the shores nearby to search for sea slugs and sea pens and other tiny marine creatures, collecting and marking unusual specimens and recording notes on the creatures inside their journal. Grant is the protégé of the famous French naturalist Jean-Baptiste

Lamarck, known for playing around with the fundamental idea of creating life through natural processes, was knowledgeable of all forms living in marine environments, such as sponges. In this way, Grant's knowledge contributed to the creation of theories and the conclusions drawn from Darwin's observations. He was the one who pointed the out Darwin the connections between primitive marine invertebrates. It also helped provide Darwin with clues as to the evolution of species.

Lamarck

The members who are members of The Plinian Society received two of Darwin's observations during the at the meeting. The first was that Darwin pointed out that the Ovum (female reproduction cell) from the marine mat for larvae, referred to in the scientific name "Flustra," a rarely-trodden area of study he explored himself, had organs that could perform motion. Then, he declared his findings that the tiny black spores found on the shells those of the Fucus loreus (a kind that is an oyster) were actually egg of the pontobdella Muricata which was a skate leech.

Darwin may have been bored with the classes offered by the university however he was awed by the variety of minds that were admitted to the university. These radical and "freethinking" students and professors with a lot of rejections and science students thrown out at Cambridge, Oxford, and other conservative Anglican universities were able to expose Darwin to a myriad of new research being released by scientists from all over the world. Darwin was a frequent participant in meetings organized through students from the Plinian or other university student associations as he listened intently and in immense awe of the radical ideas of his fellow students.

There were a number of controversial theories, including one that suggested that animals had "human mental faculties" and another that denied the existence of God or any other god, for that matter , by claiming that the structure of all living things was conceived by nature and only. There were discussions, like one about what is known as the "product of a brain material," were so taboo that they were disapproved of by the authorities who intervened, however the first time Darwin came across "radical" theories and concepts that were not conventional theories. Additionally, he experienced first-hand the social consequences that resulted of defending these views that were antithetical to the truth.

Despite the remarkable advancement Darwin made with regard to his most cherished interests, Robert was still highly unhappy with his son's low scores. Disappointed by his son's apparent unresolvable school performance and yet unwilling to give up possibility, Robert extracted the 18-year-old Darwin from Darwin's University of Edinburgh in late 1827. Darwin was somewhat irritated by the fact that he had moved twice in less than two years, but he acknowledged that it was good for him. Not only was he disgusted by blood in general, but that Darwin was not able to tolerate autopsies of humans, let alone live surgical procedures. This was a realization that came to him while he was attending the horrific procedure of a small child, without anesthesia.

In the month of October, 1827 Darwin received admission to Christ's College in Cambridge, in which he studied to become a pastor. The Wedgwoods and Darwins Darwins typically referred to their identities in the form of Unitarians and even went to Unitarian services, however the latter were reported to be leaning towards Anglicanism. Although Robert himself was said to be an open-minded freethinker He made the decision together with his wife to baptize nine-month old Charles to his Anglican St. Chad's Church in Shrewsbury. Robert was religious but was not necessarily a reverent and a church-goer. In this regard, he declared that he preferred to receive a medical degree, or at the very minimum the educational Bachelor of Arts degree from his children however, the physician who was running out of options, believed for the discipline and morality offered by the Cambridge curriculum could instill a more resilient spirit of determination and direction in his son's reckless son. As part of his application Darwin must sign "Thirty-Nine Articles" which is a similar commitment to the Anglican creed. Although Darwin was not without his number of doubts, and was not a religious person at all, he determined to put his foot down and finish his studies through to ensure the father he had fought off completely.

It is not surprising that the changes in environment did not change Darwin's behaviour, including his dislike of "mainstream" training. He only skimmed a few pages of his theology texts and spent his leisure time with his trusty horse he taken home from his parents. He also enjoyed drinking, hunting, and collecting beetles along with the sons of other squires. as did Darwin his family, the obligation to his friends attended far outweighed their actual interest in their respective majors. However, Darwin capitalized on the advanced facilities and resources available in Cambridge that allowed Darwin to pursue his interests.

In the autumn in 1829 Darwin is now an undergraduate in the 2nd grade student, discovered an advisor in the Reverend John Henslow, a popular botany professor. Darwin was the first to arrive , and the at the end of Henslow's lectures, but their connection was one that went beyond the confines of the classroom. They would often be found on campus, chatting away; Darwin was also present at the extravagant dinners Henslow held at his gorgeous home, hosted by local researchers as well as students eager to explore the minds and open into the ears of fellow students. In the following days, Darwin launched a microscopic research on pollen. It was which was managed by his botany professor. The collection of his beloved beetles continued to grow to the point that Darwin had to order an individual, six-drawer cabinet to house his ever-growing collection.

Henslow

Another fascinating extracurricular event Darwin was a huge fan of throughout his stay during his time at Christ's College was the so-called "Glutton Club." Feasting on exotic birds and other un-tried-before animals was their passion. The lack of accessibility to the majority of the most sought-after dishes, aside from bittern and hawk, hindered the culinary group from meeting their goals most times. A member of the club John Herbert elaborated, "We did not have any of us served in excess, but our dinners were modest search[sic] (sourced with no attention) and well-served. And we usually ended in the evening playing the game of light vingt-et un (an early variation in the popular card game Blackjack)." However the real purpose of the society was to compete with the fraternity that was competing with it, or as Herbert said"to "show our disdain for another group of people who identified themselves by a lengthy Greek name, which translates to "dainty," however, they lied about their claim to be a fraternity by their regular habit of eating at a roadside inn...on the mutton chops, bacon and beans."

In the end, The Glutton Club eventually disbanded when members were savaged and fell sick due to a rotten part of brown owl meat that left an disgusting, "indescribable" aftertaste in their mouths. In the late 1970s, students from British universities began to hold "Phylum feasts" every 12th day in the month of February (Darwin Day) to honor the legendary scientist. According to the custom that has expanded to other countries, admirers gathered to eat a potluck meal that was designed in order to make the meal as "biologically diverse as is possible." These menu items comprise Minke whale meat and escargots as well as smoked turkey slices "monkey gland martinis" and the cleverly named "Primordial Soup" an alcoholic drink that is typically made up of chicken stock, milk miso, tuna flakes wild woodland mushrooms, dubbed "ceps," sprouting beans seaweed biscuits, and sprouting beans.

Alongside his bizarre eating habits, and the ever-growing diversification of his collection of beetles, Darwin became growingly attached to his guns and hunting rifles. "When at Cambridge I would try to practice throwing my weapon to my shoulder with a look-glass to make sure the shot was with a perfect straightness," Darwin later recalled in his writings. "Another option was to convince a companion to hold up a lit candle, then shoot at it using the nipple's cap and, if the aim was correct, the tiny bubble of gas would blast through the candle." Darwin conducted these target exercises in the dark frequently enough that students such as Darwin, the "Tutor in his College" Darwin himself frequently felt the horse-whip-like sound of Darwin's guns when they passed by his luminously decorated and lit room.

As the day of graduation drew nearer, Darwin, who dreaded the familiar expression of displeasure on his father's faces, encouraged himself to focus on his education. To earn the Bachelor of Arts degree, it was necessary to get over the first hurdle: the "Little Go,"" an exam for the preliminary stage designed to test students' theological and classical knowledge. It also determined the likelihood that students were promoted to the next stage. Both fields seemed almost foreign to the scholastically-resistant Darwin, but a few weeks of last-minute cramming, his memory of William Paley's Evidences of Christianity, and what Greek and Latin he retained from his years at grammar school, was enough for him to sail through the exam with ease. In the middle of March 1830, an ecstatic Robert was informed that Darwin was able to pass the "Little Go,"" that was proof to the happy Robert that Darwin could do well when he put his mind to it.

In 1831 Darwin scored 10th of 178 on his annual exam at the end of the year completed his graduation at Christ's College with honors.

Chapter 2: The Voyage

Darwin as an untried man

"A man who is willing to lose an hour of his time has not yet discovered the worth to life." -- Charles Darwin, Autobiography and Selected Letters (1958), Page. 145.

After receiving his degree, Darwin utilized the ample spare time he had been awarded to continue his study of natural historical. He visited libraries in the area and read through their entire collection on natural historical events. He also purchased his top works on the subject including the paley's Natural Theology, John Herschel's Natural Philosophy, and Alexander von Humboldt's poetic personal narrative of his Travel to the Equinoctial regions that comprise the newly discovered Continent. It was this plethora of new knowledge as well as the amazing descriptions of exotic regions found in the books which fuelled his desire to explore these exotic locales.

After the graduation ceremony, Darwin began to make plans for his trip to Tenerife one of the largest and most populated among the Canary Islands, where he was planning to study and assess the diversity of species of wildlife, rock formations and other natural features in close proximity. The invitations were extended to Reverend Henslow as well as many of the most sharp aspiring biologists, botanists, zoologists and ornithologists. Entomologists were also part of the class that he graduated from. Darwin went as far as to create an itinerary, and even secured an L200 (approximately $21,940 or L21,000 USD at the time of writing) gift by his dad. Then, he spent the entire month summer in North Wales, observing and helping Professor Adam Sedgwick in his geological research. However, he was shocked to hear about a prospective group members had tragically lost his life at the sea. The following weeks, those who committed to the trip were able to leave one after the other which left Darwin with no other choice than to abandon the expedition.

The dejected Darwin returned to Shrewsbury within a short time. Being stripped of the only thing he'd been hoping to experience in some time and felt like he was floating around aimlessly. He didn't know what was to him ahead but he was committed to the likelihood that he would be appointed a rural vicarage, which would allow him to follow his naturalist interests.

In the darkest of times for Darwin the unopened letter that was awaiting Darwin in the study at Mount House was about to transform his career. After just returning back to Mount House, he was at the point of packing his bags for two weeks of shooting partridges with his family and housekeepers when one told Darwin of the letter. The excited young man rushed into the study immediately and slicing away the seal of wax in his speedy pursuit. The twinkling eyes of his were growing larger as he read the contents of the letter

- it invited him to join in the crew of HMS Beagle, where he was to act as an "resident doctor-naturalist" for a period of two years. The round-the-world expedition was funded by the Royal Navy, aimed to highlight Patagonia which is the southernmost tip of the South American islands, and was to be commanded by a seafarer aged 26 named Robert FitzRoy. The captain was Reverend Henslow who was approached by FitzRoy to provide the list of potential candidates. He endorsed Darwin.

FitzRoy

S.P. Hunter's photograph of a reproduction of the Beagle

Darwin did not waste time in accepting the invitation to ensure that the chance does not pass by him. Then, he walked into Henslow's office to talk with him on the particulars for the voyage. This was not an opportunity for a junior crew member and was a person who was of the same rank as the captain "who is afraid of loneliness in the command." Furthermore, Darwin learned the principal motives behind the voyage and that was to map and map the coastlines of Patagonia and its harbors in hopes that they would promote commerce, defending British interest, as well as increasing the knowledge of the Royal Navy about shipping routes.

FitzRoy was also planning to release three of the four Patagonian Indians he had kidnapped from Tierra del Fuego in retaliation for the theft of one his whaling vessels from the area just more than a year ago. The four hostages who he named Jemmy Button Fuegia Basket York Minster along with Boat Memory, had been treated to well-meaning but degrading social experiments over the past 15 months. All four, two of which were children, were forced to live on the margins of London and were dressed in 19th century European fashion, instructed in English social etiquette and morals and, finally, Christianized. Unfortunately, the 20-year-old Boat died of smallpox this meant that the task to plant the seed of Christ over this "dark continent" is now upon the shoulders of Jemmy, Fuegia, and York.

Before Darwin was even able to think about embarking on the ship, Darwin had to get not just the approval of, but also the promise of financial patronage from his famously rigid father, who was only to receive remuneration by the knowledge. In the beginning, Robert, who suspected that this trip was nothing more than a reason to waste Darwin's time, was adamant over Darwin's requests. There was no reason to get stoked about any "wild plan," scoffed Robert. Additionally, this kind of activity was impulsive hazardous, dangerous, and not the typical of a man from religious piety. If Darwin could "find any person with good sense who would advise himto move on," said Robert, could he rethink his decision.

When Darwin was confronted with the challenge He immediately jumped in a car and headed to the home of his uncle Josiah Wedgwood II in Staffordshire. Josiah II, an outspoken Abolitionist, the heir to the Wedgwood Pottery Company, and the future MP for Stoke-upon-Trent was considered to be one of the most trustworthy and intelligent people in England at least as when Robert could be concerned. So, when Darwin was successful in convincing his uncle to endorse his ideas - and wildly in the process the exuberant way Robert was unable to do anything other than grudgingly give Darwin his blessing.

Josiah Wedgwood II

To help him prepare for the thrilling, yet physically exhausting and mentally demanding journey, Henslow advised him to look up the newly released premier volume of Principles of Geology, written by Scottish geologist Charles Lyell. However Henslow, a conservative Anglican trained Henslow encouraged Darwin to take Lyell's ideas with the salt of a pinch as the author advocated an uniformitarian (gradual and gradual change) over the more catastrophic (short rapid, abrupt vlolent bursts of natural catastrophes) method to explain geological developments. After hearing this the Captain FitzRoy generously presented Darwin an original copy of the book that was in question. Notwithstanding Henslow's lukewarm opinions about the up-and-coming lawyer-turned-geologist, Darwin soon became a self-professed "zealous disciple of Mr. Lyell's views."

Lyell

Darwin went on to London in the afternoon, where he met with FitzRoy to the very first time, and was given basic information about equipment assembly as well as specimen collections. He told a acquaintance about the "capital enjoyment" it was having by indulgencing his self in a spree of shopping for equipment however, he cut down the list of purchases in a note to his home, while feigning thriftiness. "I am able to purchase a couple of strong and reliable pistols as well as some excellent guns for less than L50 (roughly L5,250) There is an opportunity to save money," he fibbed. "[I have also purchasedan excellent telescope with compass, (L525) and they are the cheapest equipment I'll need." He didn't refer to the defensive club the clinometer (a instrument for measuring the angles of elevations and angles) as well as the pantograph (a kind of drawing the compass) as well as a the hygrometer (an instrument that measures water vapor within the atmosphere and on earth) along with other expensive equipment he took along for the trip.

After the most brutal winter cold and the accompanying storms were gone The date was then set for the trip: 26th December 1831. The 73 crew participants of the team exhausted from the drinking and Christmas parties that they enjoyed on earlier in the night did not make it to the vessel, as it were. They were capable of setting sail at Plymouth Harbor the following morning at a time when the pain that was affecting their temples had eased. The Beagle's roster aside from FitzRoy and Darwin they were both academically and technically proficient men like 1st Lieutenant John Wickham, 2nd Lieutenant James Sulivan, Surgeon Robert McCormick and his assistant Benjamin Bynoe, and account-keeper George Rowlett.

While the diverse and highly-spirited gentlemen in the crew - who had an average Age of 25, interacted with ease, life aboard Beagle Beagle turned out to be more difficult than Darwin thought it would be. In the beginning it was a warship that had been recycled that was renovated to meet the requirements of its 73 passengers leaving Darwin shocked by the sheer size and poor comfort the Beagle offered. The cruiser weighed 242 tons and measured over 90 feet long and was built to be space-saving and flexible with an eye towards luxury. While the ship had an impressive library that contained more than 300 books but the poop deck Darwin shared with two other males at back of the vessel was not bigger than 10x11 feet with the mast taking the majority of the area. After the sun had was set, Darwin retired to the cramped quarters that he was given and was lowered onto the hammock, which was squeaky and hung over his desk. The only consolation was the soothing sounds of crash of the waves and the breathtaking sight of the night sky through the skylight that was just two feet higher and that would never fail to soothe the tired scientist to sleep each night.

The casual contact that resulted from the endless hours he was with the Beagle team also led to the development of a variety of behaviors behavior, habits, and behaviors that he'd never seen before. Many of them were not the best. One example is that Darwin was soon to realize that a flimsy belief killed his chances of being in the Beagle. It turns out that Captain FitzRoy was a huge fan of physiognomy, the millennia-old "science" known as physiognomy, which is capable of determining personality of a person based on their facial characteristics. Darwin's big, "beak-like" nose, FitzRoy initially claimed it meant the person he was talking about should be avoidedas the man was likely to be suspicious and untrustworthy, perhaps possessing criminal potential. Darwin later admitted his shock within his diary: "[FitzRoy] was an extremely devoted follower of Lavater and believed that he could determine an individual's character based on the appearance of his features He was also skeptical that someone with my nose could be able to summon the necessary strength and determination for the journey. However, I think he was subsequently satisfied that my nose was speaking falsely ..."

Darwin saw an additional glimpse into FitzRoy's meticulous manner of conduct after he was made aware of the many requests that the captain hounded the engineers of Beagle with at sea. FitzRoy was particularly observant in handling the instruments used for research aboard the vessel. In an effort to stop the original iron cannons of the ship from causing disruption to the beagle's "precision compasses" the petitioner requested them replaced with brass counterparts.

However, when his request was rejected and he had to replace them from his pockets. FitzRoy was also unsatisfied with the 12 marine chronometers (used to aid in celestial navigation as well as to determine lengthitude) which were installed on board which he added to the total number. Any person - besides FitzRoy and one other crew member who attempted to alter the ship's equipment and devices were slapped by the captain.

A chronometer that are on the Beagle

One of the most intriguing of Darwin's early concerns however, was the mild ribbing and even the occasional ridicule that he was subjected to for his apparent devotion to God and his God-like qualities, which that he had previously been unaware of. Although he seldom sought advice out of the leather bound Bible that he carried on board and was "heartily ridiculed by a number of the officers...for that he quoted the Bible as a source of unquestionable authority in a moral matter ..."

Then, and perhaps most importantly on the voyage that ran for three years more than the anticipated 2 years, the weakened Darwin was plagued by severe bouts of seasickness. Because of this and his insatiable desire to discover the unknown, he decided to spend most of his time scouting for specimens in the waters while the others on the crew were busy by surveying and measuring in the deeps of the ocean. In reality nearly 70% of a novice navigator's time was spending time on land.

Semhur's map of the voyage's routes

On the 4th of January 1832 the ship Beagle tried to dock on Madeira Island, their first stop to rest and replenish things, but a sudden windstorm prevented them from reaching shore. Darwin was not aware of that the ship was skipping over the Portuguese archipelago as the man was so exhausted and vomiting to leave his room. However, when the Beagle defeated with a strict 12 day quarantine was allowed to continue sailing over Tenerife which was the first stop-off point and the naturalist's ideal location, Darwin was devastated. "This was a major disappointment for Darwin," FitzRoy said. Darwin," FitzRoy later said. "[Darwinhad] always had a desire to visit the Peak. To be able to see it - an anchor and be near the landing, but be forced to leave without even the slightest chance of seeing Teneriffe [sic] again was to him a tragedy." However, in any case, Darwin chose not to think about the tragedy and wisely took advantage of the route through St. Jago (Santiago Island) in the Cape Verde to carry out an investigation on plankton.

The Beagle was spotted at St. Jago, roughly 300 miles from the African coast on the 16th day of January. After Darwin had collected his tools and hammers to get through the sleepy range of volcanic mounts as well as "singularly barren terrain" before finding his first tropical forest that was encased by a small gap between two hills. Then, Darwin marveled at the amazing variety of insects, plants along with sea life, particularly the vibrant group of cuttlefish that floated through the water in the tide pool. Darwin's fascination with the hilarious creature's ability to blend in with the surroundings is evident in a letter to the Reverend Henslow: "I took several specimens of an Octopus that was able to exhibit the most amazing capability of changing colors and blending into the surroundings, comparable to chamaelion and clearly adjusting in the hue of the ground it was passing over. Yellowish and dark brown with red being the predominant hues. This is believed to be a new phenomenon, so it is possible to find out.]...Geology or the invertebrate species will be my primary target throughout the entire voyage."

Joao Carvalho's photo of the cuttlefish.

In St. Jago that Darwin discovered what appeared to be among at least, the earliest indications of geological changes. Seashells that had been condensed which was once on the seabed, is now scattered across the cliffs at approximately 45 feet above the sea level. This was a sign of strengthening the rudimentary theories of Lyell of a planet slowly changing in a measurably manner as time passed, something that was not commonplace in earlier times of the Victorian Era. Yet, Darwin remained faithful to this theory to the point that Darwin began to develop his own theory about sinking ocean floors and rising continents.

The Beagle was sailed from St. Jago on the 8th February and moved westward on its way towards Brazil. Prior to arriving in the area that was once the Ilha de Vera Cruz ("Island of the True Cross") The vessel was stopped by a group of fifteen cragged, thin pieces of land that they called "St. Paul's Rocks," totaling approximately 4 acres. Finding these islands was an achievement in and of itself, as the closest continental terrain was over six21 miles. This is where FitzRoy and Darwin discovered the tiny islets were the only remnants of a once-majority mountain. Darwin was amazed: "Bottom could not be located within one millimeter of this island. If the depth of the Atlantic is as deep as is commonly believed it must be, then what a massive pyramid that must be."

John Vergan's photograph of the place

Darwin also observed bizarre chalky rocks and hillocks that protruded from the ground. They "appear from an extended distance with a stunningly transparent white hue ...[andan overcoat of shiny polished substance that has a pearlescent luster with a grayish white shade." It was only after closer examination did Darwin recognize that these deposits were semi-fossilized, hardened "dung of a huge number of sea-fowl" that the land had accumulated over the course of several millennia. Also called "guano," the substance was extremely sought-after by Brazilians and the other locals nearby because of their high levels of phosphate. It was also an efficient fertilizer, it was also an excellent source of nitrates that were used in gunpowder.

After the crew members had stocked up with food, with seagulls cooked in a campfire, and shark flesh After a few days of rest, the Beagle headed towards Brazil and crossed the equator on February 16th. Its crossing across this fictional frontier allowed Darwin up to yet another tradition, and the occasion was marked by fun-filled celebrations that gave sailors a short break from their work. In accordance with naval custom, FitzRoy, sporting - or barely sporting the dress that depicted the sword-wielding Roman ocean god Neptune invited people who were crossing the vast divide between the southern hemisphere, also known as"the "Kingdom of Neptune," for the first time. Darwin was among the first-timers who was required to undergo an elaborate ceremony to initiate them which was similar to a contemporary fraternity hazing ceremony. Veterans of the sea, "Neptune's constables," were assigned the task of making these pledges feel humiliated and hounded as well as "griffins," as they were referred to in a variety of ways. Blindfolded, Darwin was taken to an ungainly plank that was suspended over the tub of numbingly cold water. As he struggled to stay straight, the constables splattered his jaw, face, and a few of his armpits with oil paint, and pitch. They after which they scraped off his hair using the aid of a "piece of iron hoop that was roughened." In a flash, Darwin was "baptized" and plunged headfirst into the water. There, it threw up and shook until he finally wiggled to his free will.

On the 28th day of February on the 28th of February Beagle came up to the shores of Todos os Santos ("Bay of All Saints") in Salvador which is the capital of the Brazilian state of Bahia and set an anchor across the railing. The next few days the amazed Darwin strolled along the riverbanks and explored the rainforest taking in the splendors of Bahia. Darwin was amazed by "the elegantness that the grasses display, the uniqueness of the parasitical plant, the beautiful flowers, the lush green foliage ...[and the general splendor of the plants ..." from the outskirts He observed the daily eating, nesting and mating behaviors of the creatures of nature and began to think about the ways in which these different kinds of life survive and coexist with each other in a constantly changing surroundings.

Also, it was during that trip to Brazil that Darwin observed the Brazilians' brutally inhumane treatment of their African slaves, for the very first time. He was apparently so afflicted by the blatant violation in human rights, that he made his way towards the beagle and walked into the cabin of the captain and pleaded with him for solutions to this quaint situation. To Darwin's chagrin and instead of siding with his side, FitzRoy insisted that he stay away from the situation and the slaves were satisfied. Naturally, Darwin took even greater offence at the remarks of FitzRoy and as tensions increased the scuffle became so fractious that the insecure FitzRoy banned him from ever sharing a meal with him ever again. Also furious, Darwin went back to his room and began looking over his bags until a couple of days after, when a smug FitzRoy came to Darwin's door to express his apology.

A lot of people believe that Darwin's encounter with, and the resulting compassion for slaves started to erode his faith, which was not any way solid. He was unable to comprehend what a God could do to force people to endure such miserable lives. "But I was gradually coming by this time to recognize how it was the Old Testament from its manifestly false account of this world including its reference to the Tower of Babel and the sign of the rainbow, etc...and by appointing to God the emotions of an angry tyrant could not be believed than the holy texts of the Hindoos or the views of the sages of any barbarian...that the Gospels can't be proven that they were written in tandem with the events...Thus doubt sank into me at a slow pace ..."

On the 18th day of March on the 18th of March, the Beagle left the Bay of All Saints and returned to the sea and headed towards the shoals of the Abrolhos Archipelago. It is a collection comprised of five coral reef-ridden islands. While the ship sailed through these waters that were shallow crew members carried out the sounding and depth measurements. Darwin was as a contrast was able to explore the interior and, after getting acquainted with Gauchos from the area (South America's cowboys) they let him rent a modestbut cozy house near Botafogo Bay.

After settling into the new home and settled in, he began to convert ink wells after ink wells into notes about creatures, plants and geological structures within the region. In one of his initial research projects, he analyzed the tiny worm-like creatures that covered the ocean's surface with brown hints. He also tried hunting, and began to store samples for Henslow's offices back in the Motherland.

In the beginning of July in the beginning of July, the Beagle was able to glide onward towards Montevideo in Uruguay. It is the Uruguayan capital located about 1,473 miles to the south of Botafogo. Again, FitzRoy and his men looked out over the coast as sea-smart Darwin explored new discoveries on the land. The 19th August Darwin had collected enough specimens and notes to send back to Cambridge. The first batch consisted of an assortment of rocks an accumulation of dehydrated and pressurized plants as well as four containers of animals, marine mammals preserved in chemicals and spirits, as well as an abundance of beetles. All of it was labeled and cataloged meticulously.

One of Darwin's most important discoveries included fossils of extinct animals which certainly predate Biblical times. They were retrieved from the rock formations of Punta Alta in the latter part of September. These included the megatherium, an underground dwelling relative of the sloth; mastodons, the precursors to elephants; a bizarre large horse with a cup that looked like an anteater's shells as well as the skeletons of massive, rodent-like mammal. The paleontological knowledge of Darwin was a bit limited, but having realized that they were like gold to dinosaur experts back at home, he began to view them as gold, and it was even necessary for him to get the valuable goods onto the vessel since FitzRoy initially dismissed them as "useless rubbish." In reality, his observation was partly correct in that the majority of these fossils were belonging to creatures that were not known to paleontologists at the time and therefore were not much of a benefit.

The Megatherium's skeleton

Then, Darwin was understandably spurred on by the discovery of the fossil-rich gold mine and began dedicating more of his time the excavation of these ancient animals. The discovery also raised many more questions. What exactly was it that drove out the creatures? What is the process by which the creatures which now inhabit the earth develop? The more Darwin was driven to find solutions, the greater precise evidence he discovered that questioned the consensus of creationist theories. In Argentina For example, Darwin discovered that grass patches that the cattle grazed on were less dense and finer than patches left unaffected by livestock. The scientist concluded that the difference was due to stool, or feeding of cattle, suggesting that the overgrown and dense areas of land that were not inhabited resulted from natural processes and not the intervention of God.

As uncompromising an abolitionist Darwin was however, he was not perfect and held a host of reprehensible beliefs about tribal people. When he wrote his journals regarding Tierra del Fuego, the southernmost tip of South America, he repeatedly calls people from the Fuegians as "miserable and degraded savages" and also he called one particular group "the most miserable and abject creatures I have ever beheld...in the lowest quality of living than any other region of the world. ..."

Illustration of Beagle an anchoring in Tierra del Fuego in 1832

"I couldn't have imagined how vast was the gap between a man who was savage and one who was civilized," Darwin ranted. "It is much greater than the distinction the difference between a domestic and wild animal. ..." In order to not appreciate the beauty of this particular society, Darwin sneered at their "hoarse click, guttural" tongues, and made fun of their clothes: "[W]ith their naked bodies covered in black white, red and white and they resembled many demons who were fighting...The gathering was akin to the devils that appear on stage in plays such as Der Freischutz." Darwin predicted the possibility that "civilized" races will eliminate "savages" over the course of a couple of decades, however the crew members of the Beagle were foreigners living in the foreign country and were required to behave diplomatically with their hosts.

The two sides had a difficult time communicating initially and even had Jemmy or York as translators, but the tensions subsided after the people of the tribe gleefully received their roll of red cloth. The longer Darwin was with them, the more the more he realized he'd completely underestimated the intelligence of the tribe. Locals interacted with strange-looking foreigners, who they wore with their pale skin, huge eyes and scraggly beards by gestures such as the pat of a crew member's chest to signal the friendship. They Fuegians were also equipped to understand and imitate unfamiliar languages, and sometimes, reciting complete sentences to crew members in almost perfect cadence and elocution.

But Darwin's initial resentment did not turn in to respect. It turned unnecessary pity "These poor, squatters were restricted by their growth...their skins were filthy and greasy and their hair was tangled. their voices slurred, their gestures violent and lacking dignity. In the eyes of such individuals, it's hard to convince oneself that they're fellow-creatures and part of the same universe." Darwin was, however, astonished by the ease with which the indigenous people were able to adapt to their harsh environment. In the winter, they were thick-skinned Fuegians built sturdy tents and rested on the frozen ground, enduring the snow and sleet, wearing little to no clothes. Darwin said, "Nature has fitted the Fuegian to the climate and the productions of his poor country."

Darwin had a second once-in-a-lifetime moment in mid-January 1835 in the middle of January 1835, when the Beagle was anchored at Chile's harbor of Concepcion. On January 15, 1835, Chileans across the country were shocked by the sudden rumbling sound of the ground. People outside were able to crane their necks to the north, towards the volcano of Mount Osorno, the source of the mysterious sound however, it wasn't until the mountain's peak released an erupting cloud of charcoal grey smoke, which was then a threatening rumble and then they began to spread. A tidal surge and earthquake that measured 8.8 also shook the nation just one month after.

The destruction and the damages that ensued was truly painful, it also helped in advancing Darwin's theories of evolution. The seabed gravel that was deposited by the harbor was, for instance, been lifted at minimum 3 feet. Furthermore, the tidal wave released a mass of seaweeds that were knotted, shellfish and other unlucky sea creatures, scattered across dry and elevated land. Over time, he put together the link between changing environment - usually the result of massive natural catastrophes - and the disappearance of species from long ago.

In the latter half of 1835 in the last quarter of 1835, the Beagle arrived at the spot that would leave the longest lasting impression in Darwin's mind that of the "frying hot" amazing Galapagos the Galapagos Islands, a group of thirteen "volcanic prison" islands located 600 miles from South America. "[The Islands were] extremely amazing," Darwin gushed. "It appears to be an island within itself with the vast majority of its inhabitants including animal and vegetable and marine, are found nowhere else."

Darwin was fascinated by the hummingbirds as well as the massive marine iguanas, which plied through the shallows and dined on the mossy seabeds. But Darwin was especially fascinated by tortoises of enormous size that floated on the shores. the naturalists and crewmen mounted and rode as donkeys. Darwin was also fascinated by the variety of finches that resided in the Galapagos as well as how each species was found in different islands.

A marine Iguana

Particularly, the various dimensions and shapes of finches' beaks appeared to be contingent on the distinct geological characteristics of their natural habitats. Finches living in areas that were populated with fruit trees and shrubs were equipped with massive "parrot-like" beaks while those that ate insects as well as tiny invertebrates, such as maggots and earthworms were equipped with narrow pointed beaks. Darwin was right to say, "One might really fancy that, from a plethora of birds...one species was altered and adapted to various purposes."

There is a theory that the Galapagos finches triggered the light bulb inside Darwin's head. He believed that, thousands of years ago an intense wind gust caused a flock of birds to be separated away from mainland life, dumping an entirely different batch on each island in the vicinity. To be able to survive, finches had to survive on the smallest sources of food were available to them. And over time, this caused the various batches of birds to adapt to their new home and eventually branched off into an individual species. Others argue Darwin's "eureka moment" was much later.

When in the Galapagos naturalist revisited his unusual eating habits. He indulged himself in an assortment of local dishes including barbecued armadillos that was said to taste like duck. He also sampled rheas Iguanas and "buttery" tortoise carcass. Agouti's flesh "20-lb chocolate-colored rodent" was, according to him, the "very most delicious meat he'dever tasted."

Darwin took the final days of his journey hunched at his desk, working on his 770-page journal, and stacking his 1,750 pages of notes and recording the 5,436 pelts fossils and corpses Darwin had discovered. While he was doing this his mind was buzzing with new information and an new set of questions.

In all the trip that lasted for five years was a total of Darwin more than 40,000 miles around the globe, and the naturalist himself walked more than two hundred miles of tropical rainforest on feet. At the time that Darwin's Beagle came back to Plymouth in October 1836, the road to his profession had been laid and consolidated. Darwin put his rifles from hunting in trunks and moved his robes as vicars into the side in his wardrobe. He then cleared his desk and covered it with books, piles of his personal journals and notes and an assortment of jars and frames of his discoveries. He was now ready to begin with a fresh chapter.

Chapter 3: Love and Prestige

"Ignorance more often breeds confidence than knowing: it's those who are ignorant, not those who are knowledgeable that so confidently declare that the problem is not going to be solved by science." -- Charles Darwin, The Descent of Man (1871), p.3

Darwin as well as Beagle crew, were received by a crowd that was much larger than those who sent them off five years ago. Their enthusiastic followers followed Darwin, applauding the "heroic" efforts, while giving them gifts, flowers as well as a myriad of positive spirits throughout the journey. In the crowd of people pouring out their happy wishes and praises to Darwin were among those who were the highest-rated gentlemen within the realms of naturalism as well as geology. It was evident that he was a part of the group.

Darwin did not seek out praise, but he soaked up every ounce of the highly-deserved recognition he received. the majority of it came from his father Robert who presented Darwin with an L400 annual allowance, which is around L42,000/$58,258 USD in the present. This was the beginning of the golden age of his professional career.

After the return of Darwin to England, Darwin was blessed with the privilege of being introduced to his idol Charles Lyell, for the first time. In the month of January the following the year Darwin became a member of the Royal Geological Society, wowing his fellow members with his opinions regarding and the potential ramifications of the expanding Chilean coastline. In the early 1838s the lionized naturalist was elevated to the position of secretary for the Society. He would not stop trying to climb the ladder of fame from there.

In the middle of June that identical year Darwin had been invited by the Athenaeum Club in late June of that year. be a member of in the Athenaeum Club. The Athenaeum was an exclusive London-based society that was founded 14 years before to Darwin's arrival, and was open for only reputable academics - male and female , in the upper levels of society who had attained the distinction of a notable academic regardless of whether it was in the fields of literature, engineering or even science. The only members that included Charles Dickens, Humphry Davy as well as Michael Faraday, amongst others were able to take advantage of the elegant and lavish amenities that were available in its Neoclassical clubhouse.

Darwin made the most of his privileges as a member. And starting in August the clubhouse became his regular stop. the clubhouse on a daily basis to peruse the thousands of titles available in the library's vast collection, had a meal at the "Coffee Room," snuck off for a smoke within the "Smoking Room," and took naps inside one of the luxurious suites. Later, he recalled, "[I] dined at the Athenaeum as a gentleman or, more accurately, like a Lord, because I'm sure that the first night I was in that grand drawing room, seated on my own on a couch I was akin to an duke...I am awestruck in the Athenaeum where one can meet an abundance of people there who one would like to see...I am enjoying it more since I was expecting to hate it."

The majority of Darwin's biographers agree that it was between 1837-1838 that the naturalist's historical theory started to take shape. In the months between, Darwin went around the country to talk to experts on the significance of his South American specimens. A discussion With Professor. Richard Owen from the Royal College of Surgeons revealed that the skull that Darwin took off the bank of the Uruguay River belonged to a Toxodon which was the precursor to the capybara, which was as large as the adult size of a hippopotamus. The evidence-based remains and the ancestors of the anteaters, sloths and armadillos found in across the South American plains, demonstrated that the extinct animals were replaced by their contemporary counterparts in the then obscure "law of succession."

Richard Owen's drawing of the Toxodon Skeleton

Owen

The next step was when Darwin went to fowl expert John Gould from the Zoological Society. Gould confirmed Darwin's theories that the birds of the Galapagos weren't actually mixed with finches, wrens, and warblers but rather different kinds of finches that were ground-based. The ornithologist also identified three species of mockingbirds living in various islands, further confirming Darwin's theory of building.

On top of the comparably receptive and scientifically-minded Erasmus, Darwin had a close clique of flexible friends at his disposal to run his ideas by. The majority of "dissidents'" daily dinner events were devoted to dissecting and thinking of ways to improve Darwin's theories. It was perhaps Erasmus's unwavering devotion to his brother that made Darwin decided to stick with his pivotaland fragile theory for so long. It is true that the theory was

Erasmus who saw the resemblance in Darwin's ideas , and pointed him to the work of their grandfathers, Erasmus Sr., and Robert Edmond Grant.

Darwin could have collated his notes and turned out a book by subsequent year. But he discovered himself stifled by the norms of society. Any person who spoke of an idea that even suggested a different explanation for the universe's beginnings that excluded God was instantly blasted and condemned by Cambridge religious leaders for being unbiblical. According to the legend the moment Darwin introduced the idea of apes being part of the question, the room exploded with a chorus of condescending laughter and a lot of people mocked the scientist because he was closely related to the monkey.

A caricature that depicts Darwin as an Ape

In fear of losing his hard-earned reputation and being an unpopular figure in the scientific community, Darwin was forced to hide in the dark and continue his research in secret for the remaining two decades. The pile of journals he stuffed in shows the awe-inspiring quantity of sweat as well as the amount of time that he put into this time. Drawings of family tree organisms For instance, they were jammed between the passages and in some of the notes' margins. He looked for and eliminated possible other origin explanations as well as causes for the disappearance.

The pressure of it all caused him to feel so much stress that, by the end of September 1837, it had taken its detrimental effect in his overall health. The stomach flu, nausea, and heart palpitations made him bedridden for several weeks. In the hope of easing his frequent issues, Darwin hiked up to the Highlands in Scotland in the early part of 1838, because they were famous for their healing scenery. In the Highlands, Darwin rented a quiet room to do some leisurely reading about the natural world. Darwin also made short excursions towards some of "ice-dammed beaches of the proglacial lakes" also known as"the "Parallel Routes in Glen Roy," which

Darwin compared to lifted Chilean beaches.

In the beginning, it appeared that Darwin was able to overcome his illness However, his burning desire to share his research and his contradictory desire to be respected and relevant was a constant source of irritation as his ailments came back. "The amazing fabric of the universe teeters and falls" Darwin lamented. Even more alarming, the protests and censorship of evolutionary materialism and philosophical materialists appeared to be even more insistent during the time of Scotland. Scottish capital. The names of those suspected , without evidence, of such crimes were ruthlessly detained and charged with conspiracy subverting the social order.

The end of the day, when Darwin was, along with many of the educated people, frightened about the idea of destroying Christianity Christian Bible, their itch to inform the masses was much stronger. While humanity was always in the background, Darwin now struggled with human beings' place within their place in his "tree of existence." To try and find hope to dispel the mystery Darwin went to the local zoos and observed the human-like behavior of orangutans, recording their social behavior. He also visited a variety of breeders of animals and listened to their brains on the strategies they used to create domesticated animals like horses, dogs and the pigeons. According to his journal at the period, Darwin's belief in God was waning. The belief system of society God was, according to Darwin is nothing more than a means of coping that is used by tribalists. Although he believed in this notion however, he was very aware of his position. "The worship of God [is an] outcome of the brain's] structure" he wrote. "Oh your Materialist!"

It was not until the month of September in 1838 that Darwin discovered the final part of the puzzle. In that month, he snared an original copy of Essay about the Principle of Population, written by the famous economics professor Thomas Malthus.

This essay was a celebration of the wild Victorian "Workhouse age" that was prevalent at the time and argued that regardless of the time of year there will have more hungry mouths than resources. This phenomenon is best summed up by this timeless law of life: "[When] the population increases geometrically...food production rises arithmetically...[therefore], charity is useless."

In an effort that was desperately needed to halt or at least slow down, the unruly growth of the population and increase the number of children born, The Whig Party enforced the 1834 Mathulsian Poor Law. The law was a controversial way of controlling population growth that saw physically and mentally impaired paupers were detained and taken to workhouses deliberately segregating men from women in order to stop their procreation. This logical reasoning enabled Darwin to come to this conclusion. the explosion of population like the one encountered by the English caused an ensuing struggle for resources. Only the most powerful and innovative humans won the resulting competition as the natural assets were constructed on the basis of the "first come first served" basis, which meant that they picked the poor - as well as those who were "weak," for a for lack of a better phrase each time. The same principle was referred to as "natural selection" did not just apply on humans. It also applied also to all living creatures on earth.

Darwin's final revelation earned him some admiration from Erasmus and others in his entourage, but the scientist was still at odds with himself regarding the revelation that the theory was revolutionary. However Darwin fought against disappearing into obscurity by publishing one book after the next, including Journal of Researches into the Geology and Natural Hsistory of the Various Countries visited by H.M.S. Beagle, as along with the structure and distribution of Coral Reefs and Volcanic Islands. Later, he received a grant worth L1,000 (I 105,000/$145,645.50 USD) from the city of Cambridge, and the funds were used to employ a group of experts that would analyze and draw up descriptions of his untouched specimens. They were later compiled into a single book, titled "The Zoology of the Voyage of H.M.S. Beagle.

Other than his work that was never published, Darwin had little, in fact, that he could complain about. Darwin was well-nourished - to the point that he was beginning form a somewhat of a stomach and was not averse to anything in terms of the material and physical comforts. His popularity was always rising. However, he couldn't admit that something was missing from his, something he believed that only companionship could be able to fill.

After the completion of the "transmutation" theories, Darwin took a fleeting time out from his busy life to seek married. He was only in one relationship of serious romance with a different woman, and it was brief and a few years back. The first object of his admiration was a raven-haired doe-eyed beauty called Fanny.

Fanny was who was the sibling of his former grammar classmate William Owen, is described as a soft-spoken but friendly and spirited free spirit who helped Darwin to cope with the passing from his own mother. The two had their first teenage years practically joined at the hip riding around on their beautifully groomed horses and eating wild strawberries while they chatted away the day. The enthralled Darwin expressed his admiration to her through a note written to one of his acquaintances. "Fanny who, as everyone in the people know is the most beautiful and most beautiful individual that Shropshire has, and Birmingham as well."

But, Darwin could never ignore his passion as a scientist, and Fanny would not have wanted Darwin to ignore his calling. When Darwin was immersed with his research, the lovers were further separated. Darwin continued to long for Fanny for many years afterward however any hopes of the possibility of a reunion was eliminated after Fanny took up the proposal to marry Robert Myddleton Biddulph who was a wealthy aristocrat and aristocrat, even as Darwin was from the Beagle. The heart of Darwin was broken but the note of encouragement she left in her letter to him was solid enough to hold it together that it remained in place for a period of time. She wrote to him "Believe that I am Charles I am sure that not a changes in your name or status could ever affect or diminish the feeling of affection and love I've had for a long time to you."

Darwin who was just one year away from turning 30 was still wary about committing to an affair, as the fear that it could affect his lucrative career. However, Darwin was genuinely in a state of loneliness. "My God," reads one of his journal entries dating from July of 1838. "It is infuriating to imagine spending all of your life like an uncaring bee at work, but doing nothing to do at all...Imagine spending your entire day just by yourself in a dirty and smoky London House...Only imagine yourself as an attractive, soft woman at home on a couch with a warm fire, books, and even music perhaps...Compare this image with the dim real world of Grt. Marlbro' St."

Always a scientist, Darwin even put together an entire list of advantages and disadvantages that advocated for and against marriage. On the plus side, he wrote, "Children (if it please God)...constant companion (and friend in old age)...object to be beloved and played with (better than a dog, anyhow)...Someone to take care of house...Charms of music and female chit-chat..." Negatives, on the other hand, included "[no] freedom to go where one liked...or conversation of clever men at clubs...[will be] forced to visit relatives, and to bend in every trifle...expense[s] and anxiety of children...quarreling...fatness and idleness...less money for books..." Most daunting of all to Darwin was the potential "loss of time." Regardless of the length of his cons list, by the end of his calculations, he had concluded, "Marry - Marry - Marry Q.E.D." The initials, which stand for "quod erat demonstrandum," is a Latin phrase often used at the end of mathematical or scientific proofs.

Darwin's list of potential brides was reduced to one final choice: Emma Wedgwood, his first cousin through the maternal uncle of his. The intimate nature of their relationship wasn't much of an issue for Darwin, since generations of intermarrying between their families was already established and such marriages were not uncommon in the day. He was concerned about Emma's unshakeable Christian faith and how her cousin would react to her beliefs on creationism that were not his own.

Emma

In a moment of despair, Darwin confronted the issue to his father, whom he then told him to "conceal his concerns." Darwin was well aware of the benefits of being quiet about his opinions however, after thinking the issue over a few times issues, he believed it was immoral to begin a marriage without the clean new slate. Therefore, he shared with

Emma his truthful opinions and his growing distancing from his faith. Much to his delight (and happiness), Emma, who was impressed by his honesty, agreed to his suggestion. But she was clear right from the beginning that their "opinions on the most crucial issue [the real origins of the universe" will always "differ significantly."

On the 29th day of January, the elegantly dressed groom and his beautiful bride, dressed in a stunning emerald gown made from silk, exchanged vows. The man who conceived the idea that there was "natural selection" was later to have 10 children, including the first of his cousins. Three years later Darwin and Emma have moved their group to a white-bricked gray-roofed house just one quarter mile to the south of Downe. There, Darwin revised his work schedule to include time with his children and wife in his routine. After a walk around the neighborhood each day after which he ate a quick breakfast in the company of his children. It was followed by one hour of reading letters, research (recited to his by Emma while he sat on the couch) as well as more research and a second walk around in the neighborhood with his white dog Polly. Then he popped into his greenhouse and took an inspection of the developments of his plant experiments before heading back home to eat lunch with his family. Further research was carried out in the afternoon, and threw by breaks. After the meal, it was time to play two backgammon games with Emma before retiring to bed around 10:30. Although the old man wanted to go on a trip but he was able to find a lot of comfort in the security of his new lifestyle.

A photograph from 1842 depicting Darwin and his grandson, William Erasmus

Mario Modesto's image of Darwin's study

It was apparent that Darwin was conscious of the dangers associated with procreation in a sexual relationship. Actually, he told a acquaintance about his kids being "not particularly robust." After the first birth of his child William Erasmus Darwin, the naturalist, like he was conducting a different experiment took note of every movement of his child, recording every sneeze, yawn and stretch as well as the duration and decibels of his screams during one week during his first year of life. The results were reported in an Oxford sponsored Mind journal in 1877.

Darwin and Emma mourned the loss of three of their children. However, it was the loss of their daughter, aged 10 years old, Anne 1851 due tuberculosis-related complications which truly broke him. While the loss of Anne only served to strengthen Emma's faith, Darwin felt deceived and many believe that the death of his daughter caused him to leave the religion he believed in to pursue a better life. Emma continued to take children to church on Sunday however, Darwin was confined to his bedroom, soaking up his sorrow. Anne's death Anne could have a negative impact on the bond between them but they fought back. Following Anne's passing, Charles awoke to a note from Emma that read "You should remember that you are my primary treasure (and always will be)." Additionally the conflicting beliefs might have created a divide between them but they eventually learned to respect and accept each different's beliefs.

Anne

On the 30th of November 1853 Darwin got his Royal Medal from the Royal Society for his 3-volume chronology of his geological discoveries on the expedition in addition to his ongoing research on the cirrepedia (barnacles) which was hailed as an "key part of Darwin's theory on speciation." It was a highly coveted award that was given only to 3 people per calendar year, for "contributions to our understanding of nature."

In a state of panic over the rising star of his brother, Erasmus continued to pressure Darwin to write what he believed was Charles's masterpiece. It was a wonder no one had yet dived into his revolutionary theory.

And, oh my! Erasmus was not without cause to be worried. In the early 1850s an aspirant naturalist named Alfred Russel Wallace set out to find a solution to this exact problem. In the midst of the decade Wallace published an essay titled "On the Law That Has been regulating the introduction of new Species." The work sheds a simple, but promising perspective on the evolution process. Then, a few months later Wallace who was stationed at Gilolo Island (now in the Indonesian provincial of Halmahera) was inspired by his study of the population that provided evidence of the "struggle to exist." Instead of trying to drag the issue, Wallace drew up another essay, this one titled "On the tendency of species to Continue to Separate from the original Type." He sent the essay to Darwin in a matter of minutes.

Wallace

Some critics argue that Darwin didn't discuss the letter with anyone else for a period of two weeks, because he was taking ideas from Wallace's essay which he later incorporated to his own writing. After he disclosed the contents of the piece to his close circle and his associates, they assisted in organising and putting his notes into one cohesive and digestible document. They then submitted their own essay, depriving Wallace of his rightful credit. However, others claim that Wallace did not intend to challenge Darwin's ideas and instead they worked together to improve Darwin's theories, which were later published after Wallace's approval. Darwin sent a letter to Lyell in a note "I prefer to destroy my entire work than Wallace or any other person believe that I behaved with a lack of a sense."

Wallace believed to be content in knowing that he come to similar conclusions as Darwin was, and was content to be the inspiration that drove Darwin to publish the work he considered his greatest achievement, On the Origin of Species in 1859. Wallace told his son "I have a small part in this work," after having come across the same concept that the work is based on and was referred to in the words of Mr. D 'Natural Selection, and then told him about it before the publication of the book. ..."

Final Year

"In the long history of humanity (and animals, too), those who were able to work together and experiment the most successfully have triumphed." This quote is attributed to Charles Darwin

Darwin continued to earn many more honors that highlighted his dazzling career during the final years in his career. In 1859, Darwin was awarded the Wallaston Medal from the Royal Geological Society of London. On the 30th November five years later, Darwin received an award that was among the highest-quality he'd receive throughout his life that was the Copley Medal. The silver medallion, believed to represent the "highest distinction given on the Royal Society," was an ideal venue for drama. There was a lot of drama. Church of England, along as the top participants of Royal Fellows of the

Society protested against the board's decision as they had hoped that the award would be presented to Adam Sedgwick who was who was a British priest scientist, geologist, and mentor to Darwin. The furore only intensified after the authorities decided to publicly denounce On the Origin of Species.

Darwin in 1868.

Darwin in 1878.

While Darwin's health was declining slowly. He was suffering from a range health issues as he got older like eczema extreme vertigo, boils, the twitching of joints, and general discomfort. As he grew older and the stomach-related ailments that plagued him throughout his life had largely subsided however his memory started to decline. In January of 1882, this old man was struck by an acute asthma attack as well as a dizzy spell, which was never cured from. The 73-year-old passed away on April 19th three months later due to heart issues.

Darwin in 1881.

Stanislav Kozlovskiy's image of Darwin's grave in Westminster Abbey

After the time of his passing, a passionate British evangelist , who went by his name Elizabeth Cotton, also known as "Lady Hope,"" told a group of women and young men in an institution established by American Dwight Moody in Massachusetts that she was summoned to the bed of the famous Charles Darwin. The man who was dying she claimed was studying a passage in the Epistle for the Hebrews she told him at the end of his life: "How I wish I did not have expressed my view of evolution the way I have done." He then entrusted her with assembling the followers for him in order that the man "[wantedto] tell the people of Christ Jesus and his salvation." Being in a state that there was a fervent desire to savor the blissful anticipation of heaven. ..." Lady Hope's story was later published within The Boston Watchman Examiner, and the story was reprinted time with a reprint the month of October in 1955.

Lady Hope

Darwin's family members claim that the story is nothing less than fake, written by those who wanted to trick unsuspecting people into believing in religion. As per his sons, Darwin had spent his final moments talking to Emma his wife for 43 years. He told her "I I am not in the least terrified of dying. Keep in mind how wonderful a wife you've been to me. I want my children to be reminded of how wonderful they've been to me. ..."

The conflict between Darwin's research and Christian convictions will be at the heart of his legacy, however the now famous naturalist, often described to be one of the top atheist figures in time, didn't identify as the latter. While he was adamant about organized religion , and the Christian religion in general He was an individual of the spiritual realm who bordered in the direction of being atheist. According to him, There was absolutely no chance such a "wonderful universe" was created by "brute forces." "I could claim that the inability of believing that this vast and beautiful universe, together with consciousness, was created from chance, appears to me to be the primary argument in favor of the existence of God However, how much of this argument is worth considering I've never been in a position to decide...the mind is still aching to know where it came from and the way it came about. Also, I cannot ignore the difficulty of the vast number of people suffering throughout the world...The most reliable conclusion appears to be to believe that the topic is beyond the realm of human understanding. ..."

Chapter 4: CAMBRIDGE 1828-1831.

After two days in Edinburgh my father realized that I didn't wish to become a doctor or the knowledge he gleaned about my siblings, and so the suggestion was that I become a priest. He was strongly opposed to me becoming an inactive sportsperson and that was an ideal choice for me. I paused for a long time to think because I was cautious to say that I adhered to all the teachings that were taught by the Church of England; otherwise I would have preferred to think of being a country clergyman.I was very cautious when I read in the following book "Pearson on the Creed as well as a few other books on God's Word. I reassured myself that the Creed should be fully embraced as I didn't, even think about the absolute and literal authority of any word found in the Bible.

In light of how ferociously the Orthodox was a threat to me, it is a bit shocking to think that I've always wanted to be a

A clergyman seems odd. My father's thoughts or desires, until they were discarded however, a natural demise occurred as I was a naturalist "Beagle" as I quit Cambridge. I'm in a good position to be a priest at a certain point if psychologists believe their claims. Secretaries of an German psychological society sincerely requested of me a few years ago, via letter, to take a picture of me; later, during one of their meetings was held, it appeared that the form of my face was the subject of a public discussion and one person proclaimed that I was reverent enough to have the head bump of ten priests made.

It was decided that I would be a priest and was required to attend at least one among the English universities and then graduate from them. However because I'd not opened a classic book before leaving the church, I was disappointed that I'd lost all I knew in addition to the smallest of Greek letters in the years of intermediate, as bizarre as it sounds. Thus, when the time was set at the end of the month, in

November I didn't go to Cambridge I was an individual teacher in Shrewsbury and, after the Christmas holidays in 1828, I was sent to Cambridge. I quickly restored my level of education, and could read with moderate facilities basic Greek books, such as Homer along with the Greek

Testament.

For the three years that I was in Cambridge I was unable to make up my time frequently in Edinburgh and at school for academic studies. In 1828, during the summer, I also tried maths, had a private teacher to Barmouth (a extremely boring person) however, very slowly I was able to master. It was a very unpleasant task for me, especially since I was unable to see the logic of the early algebra measurements. The impatience I displayed was reckless and I was deeply disappointed for years afterwards having not at a minimum, if not enough, to master any of the most important mathematical concepts. It appears to have an additional sense for those with this kind of talent. However, I don't believe I would have had the ability to compete at all, not even at a low-level. I took a couple of mandatory college classes that were geared to classics. The participation was hardly. I had to put in for a few months to earn the Little-Go degree at the end of my sophomore year, and I was able to do quickly. Last year, I was extremely focused on finishing my B.A. graduation and was able to brush my Classics by tackling a bit of Algebra and Euclid, which provided me with lots of enjoyment, similar to what they taught in college. In order to collect the B.A. Exploration, Paley's "Proof of Christianity' as well as his 'Political Philosophy were both required and were executed thoroughly and I think all of the evidence should be written with complete accuracy, however, not naturally written in a paley-like language. I was also awed by the reasoning behind the book as it is possible to say that I was enthralled by Euclid's Natural Theology.' The study of these works was the sole aspect of the academic curriculum that I was then and believed was a little cautious, not trying to grasp every detail of the rot.In the

The preparation of my mind it was not of any useful for me. The point was that I didn't even bother with Paley's theories, and once I accepted them with confidence, the lengthy argument that captivated me and convinced me. I've earned a respectable place among the oi polloi or the crowd of men who aren't averse to honor, for responding nicely to the test questions in Paley by learning the Euclid correctly and not failing in the classical. I'm not sure how high I was but I'm sure I did and the title of the list changes between seventh, tenth or the twelfth. (10th in the list for January 1831)

Lectures are offered in various parts of the University as well as a free of charge, but I got so bored of the lectures in Edinburgh and I didn't even go to the lively and fascinating events of Sedgwick. If I had done then, I should have become a geologist prior to my time. But I did attend

Henslow's talks on botany, and I thoroughly enjoyed them due to their consistency and amazing diagrams. Henslow was known to take his students, which included some older university students, on walks, field trips or taking rail excursions, to remote areas, or on a riverboat, as well as to the most rare creatures and plants he observed. The vacations were fun.

While my experience at Cambridge was a complete waste and, even more than wasted as we'll see later that there were positive aspects of my life. I was a part of a sporting set together with a group of lost young men, as a result of my love of hunting and shooting, and, when not so, for a ride through the countryside. We usually ate with each other in the evening, however there were times when we had people who had higher marks on dinners, and we often were drinking too much, dancing and playing cards later. I'm sure I'm ashamed of the nights and days spent that way, but because certain of my friends were really nice and we were in top minds, and with great joy I find myself looking back to this day.

It's a blessing for the fact that I have lots of other friends of completely different kind. I had a great relationship with Whitley who later served as a senior wrangler, and we went on many a stroll together. I was extremely familiar with the canons of Durham and Whitley was a former reader of Natural Philosophy, University of Durham. He showed me a taste of his photographs and lovely gravings that I bought several. I also visited the Gallery of Fitzwilliam. My taste was robust due to the best photographs I could possibly see.

appreciated. That I talked to the curator who was old. I read the biography by Sir Joshua Reynolds with a lot of interest. The flavor lasted for several years, even though it was it wasn't for me I was also awed by a few of the photos at the National Galery in London, that of Sebastian del Pombo which was thrilling to me. I also played in a musical set that was written by Herbert (Late John Maurice

Herbert who was The County Court Justice in Cardiff as well as The Circuit of Monmouth), my friendly frient whom I graduated from with a level of rivalry. I've developed a great taste in music after entering and watching these musicians play and have often taken my walks to listen to the singing in the Chapel at King's College on weekdays. It made me very happy and I would occasionally feel a tingle in my backbone.In this manner I'm sure that there was no contact or merely imitation, since I was a student at King's University by myself and often employed members of choir to perform in my room. However, since I'm totally deaf that I am unable to discern the discord, or hold time or be able to hum correctly. It's amazing what I could have done to be in a position to enjoy music.

My musical peers quickly spotted my situation and were often interested in having a look to see the number of melodies I remembered when played more quickly or slower than usual. "God help the King" was a difficult puzzle when played that way. There was a man who had a similar ear, which is odd considering that he was playing the flute. One time, during the course of one of our music exams I came out victorious.

However, Cambridge was not able to do to do anything with any enthusiasm, or brought me as the same pleasure as I snatched beetles. I was unable to distinguish them and sometimes the exterior characteristics of its species were compared with written descriptions and it was named nonetheless. Let me send you an example of my enthusiasm that one day I saw two rare beetles with one hand cutting through old bark. Then I was able to see a third fresh one, which I could not bear to lose. I put in my mouth the beetle I held in my left hand. Wow! Wow! It released an extremely acidic solvent that ignited my tongue and caused me to vomit out the insect. This was the case with the second. In the winter, I employed an employee to scrape the old trees and then put them in a huge bags. I've utilized the scraps left from the bottom of fences through which the reeds were transported, and I have a handful of uncommon specimens. I experienced lots of success collecting and I developed two different methods. The poets of the past have never was privileged to have the first poem he wrote in Stephens"Illustrations Of British Insects,' "captured by C. Darwin, Esq." Second cousin W brought me into the field of entomology. A smart guy who was at Christ's College, Darwin Fox,And I became acquainted with Darwin Fox. Then I got to know Albert's way of trinity. He became an archaeologist of repute as the years progressed and H and gathered. He was a notable farmer, chairman and member of the Parliament on a large train. Thompson was a student at is from the exact same College. A desire to catch beetles, therefore, is an indication of future success!

It is a shock to me that a lot of the beetles I saw during my time in Cambridge gave me a lasting memory. I remember the exact appearance of the beetle that I snapped an amazing photo of those pillars and ancient branches and the banks. The stunning Panagaeus crux Major was a treasure during the days before and I even observed a beetle scurrying across during a stroll in the fields downstream and after grabbing the animal, it was able to see that it was more distinct than P. crux Major and P.Quadripunctatus that is a variant from closely-related species that only is slightly different from. When I was a kid I've never had the pleasure of seeing Licinus alive and it is not a solitary thing in my eyes, however, my boys have found the specimen in this area and I noticed it was new to me. For the last 20 years, I've not seen any Englandic beetle.

It is not yet identified which has more than anyone other thing impacted my career. This was the case with Professor Henslow. I had heard of the professor from my brother prior to my the time I arrived in Cambridge as an individual who was knowledgeable of all fields of science and was therefore capable of respecting the man. He would be available at least on a weekly basis, and each student and a handful of more senior students keen on research met in the evening. I received an invitation shortly via Fox and often visited there. I was extremely familiar with Henslow in the past and spent the majority of my time at Cambridge with him in the last half of my stay, only to be called by a few of the dons "the person who walks alongside Henslow;" and in the evening, I was invited to attend his dinner with his family. His knowledge of botany, mineralogy, chemistry and entomology and geology was exemplary. His best taste was in drawing conclusions from tiny insights that spanned many years. His sense of judgment was firm and his brain was sharp.

Well-balanced, however, I believe nobody can say that he was a genius of any kind.

He was extremely religious and so traditional that he told me that even if one word in the 39 Documents was changed the way he saw it, he would feel sad in the future. In all aspects his intellectual abilities were outstanding. I have never seen anyone who thought less of themselves or their personal interests. He was devoid of any hint of pride or tiny emotion. His temper was serenely calm which was one of the most splendid and elegant of all. Yet, his temper could be sparked to a state of rage and rapid reaction by an unwise act like I witnessed.

One time I witnessed a scene that was as horrific as one was expecting to see following during the French Revolution in his company in the streets of Cambridge. Two snappers were detained as well as a mob of rougher men that dragged on the hard slippery path on their knees taken away by the police during detention. They were covered with mud, from top the toe, and their faces blowing out either because of being kicked or the stones. They appeared like dead bodies, however they were so crowded that I got only some brief glimpses into the suffering animals. I've never seen a face as swollen and angry in my entire life, as Henslow was in this horrifying scene. He was trying to get inside the gang repeatedly however it proved impossible. He then approached the mayor, and told that I should not pursue him however, he asked for additional police officers. I did not remember the incident and was unaware it was true that both men had never been executed in prison.

As the numerous outstanding projects to help his poor parishioners proved the extent of his generosity, Henslov's kindness was limitless in his commitment to the people of Hitsham for many years to follow. I would have loved to be in contact with someone like him and I am sure that Henslov was a valuable asset. I'm not able to resist the urge to share an incident of trivial significance that demonstrated its compassion. While I was studying on an un-damp surface, with pollen grains, i saw the tube exerting and was able to rush to inform him of my amazing discovery. In all likelihood, no other botany professor would've been able to resist laughing at my speed to convey such a way. However, I do agree that it was fascinating and the professor explains his significance however, he also makes it clear that it's well-known. I didn't have the slightest regret, and I was happy to discover this amazing discovery for myself. However, I chose not to share my findings in such a rush.

The Dr. Whewell was one of the oldest and most famous men who travelled to Henslow occasionally and I went with him many times throughout the night. He was the most eloquent speaker on serious subjects I have ever listened to in the company of Sir J. Mackintosh. He's written an extensive collection of papers, the most particularly Zoological.) and often remained in Henslow with his friend. Leonard Jenyns (The well-known Soame Jenyns was a cousin of Mr. Jenyn's father) and later wrote several good Natural History essays (the fish used in"the "Beagle" Zoology was currently Blomefield).I visited his home in his Fens border parsonage and took numerous walks and discussions about the natural world. I was extremely pleased with his writing. I was invited to visit. I also met a few men who were older than me, who weren't particularly fascinated by science but were Henslow acquaintances. One of the men was Scotchman Sir Alexandre Ramsay's brother , and Jesus University teacher: he was a charming man, but was living for a long time. He was a second. Dawes was a second,Then the Dean of Hereford was famous for his accomplishments in the field of low education. Along with Henslow as well as the men and others with similar standing often took trips to distant places to the country I was allowed to join and were extremely friendly.

In retrospect, I believe that something was in me that was a bit more than the typical youth group, or else the guys I've mentioned wouldn't have let me to be a part of their group as they were much older than I was and more advanced in academics. Of of course, I didn't realize that there was a higher standard and I still remember one of my friends from sports,

Turner, who saw me playing with my beets and telling me that I must to become an Royal Society fellow some day but it was a bit absurd to me.

The last time I was in Cambridge I was reading Humboldt's "personal narrative" with great concern. The work as well as the "introduction to philosophical philosophy" of Sir J. Herschel ignited in me a burning passion in order to contribute to the noble science of natural science the most important contribution. The books I have read have not impressed me as much than these two. I copied lengthy passages on Tenerife by Humboldt to read out loud during one of the mentioned excursions towards Henslow, Ramsay and Dawes as I had talked about the glory of Tenerife at a previous time and a few of the other group members claimed they would attempt to visit, however, I think they've taken it only half serious. However, I was serious and was introduced to an London merchant to inquire regarding ships. But the idea was thrown over my head when I went on the "Beagle tour.

I was allowed to choose beets to read, and also to travel around my summer vacation. In the fall, I spent all day shooting, especially at Woodhouse Maer and Woodhouse. Maer and, occasionally, with the new Eyton who was Eyton. In my time at Cambridge I had the most joy of my joyful life. I was in great shape as well, and was almost feeling good.

When I first arrived at Cambridge around Christmas time I was required to write my final exam two terms in the year 1831. Henslow convinced me to start my geological studies. Thus, I read sections and sketched out a map of Shrewsbury when I returned to Shropshire. The beginning of August Professor Sedgwick made a decision to visit Northern Wales to conduct his famous geological research between the rocks of ancient times, and He invited Henslow to convince me to accompany his expedition. (My father had told Sedgwick about the trip:

A few hours earlier they were about to walk for one mile or two at which point

Sedgwick suddenly stopped, and they promised to be to their destination, if the 'disgusting scoundrel" (waiter) had not offered the six-pence to his lady chambermaid. Sedgwick was persuaded to abandon the scene as there was no evidence to believe that the waiter was guilty of any particular perfidy.--F.D.. Then he went to bed at the home that my mother-in-law had. My mind was captivated by a quick chat with him that evening. While examining the old gravel pit in the vicinity of Shrewsbury one of the workers said that there was a huge worn tropical volte hull within it. It can be seen on cottage chimneies. I was convinced that he had actually discovered this inside the pit because the man was not planning to market it. I confronted Sedgwick about this fact that he had found it, and he declared (without any doubt at all) that someone ought to throw him in the container. However, Sedgwick says that it would cause geologically the most disappointing if it was actually embedded there, since it would rewrite everything we know about Midland counties' superficial deposits. These beds were formed during the Ice Age and I've found broken Arctic shells inside the beds after a few years.

Then I was amazed to discover that Sedgwick did not like this amazing fact that a tropical shell can be located in central England close to the surface. There has never been a moment that given me the insight that science is about organizing data, even when I've read many scientific papers, in order to get general principles or theories. Llangollen, Conway, Bangor and Capel Curig began next morning. The tour was designed to teach me how to study the country's geology. In many instances, Sedgwick directed me to the same line as his and suggested that I return examples of rocks and then mark the layers on an image. I'm not sure if he did this to benefit me since I was unaware of helping him. The tour I was on gave me an incredible illustration that showed how simple it can be to overlook seemingly obvious phenomena before they are noticed by anyone specifically. We stayed for a while at Cwm Idwal, gazing at every rock in the area with great care in the same way that Sedgwick was keen to discover fossils within them, but we did not see any sign of the incredible glacier around.

The blocks and the boulders perched along with the lateral and final moraines were all observed but that we didn't notice. But these phanomens were so obvious that a home that was destroyed by fire didn't disclose its cause in a report written several times later on the 'Philosophical journal in 1842. The phenomenon could have been more obscure as they are today were it not for the fact that it was covered by an iceberg.

The day I left Sedgwick to go on a journey through Capel Curig, walked straight through the mountains with using a compass and map, then headed to Barmouth and never took any other route, regardless of whether my route coincided. I was then in some odd places in the wild , and I loved this journey. I went to

Barmouth to meet friends from Cambridge who were studying and then went back for shooting towards Shrewsbury and Maer because I was deemed to be insane to not take part in the initial days of shooting partridges in geology and other fields of science.

"VOYAGE of the "BEAGLE" From December 27th, 1831 UNTIL 2 OCTOBER 1836."

As I returned to my home after my short North Wales geological journey, I contacted Henslow for me to inform me that Captain FitzRoy was willing to give a portion of his house to anyone who wanted to be willing to take the "Beagle" journey along with him, without compensation. Then in my MS. Journal I gave an account of the events that were taking place. I will only say that I was prepared to think about the offer right away however my father resisted with a strong voice, adding the words, "Good luck for me "If it is possible to find anyone with common wisdom who suggests you go, I'll grant my permission." The following day, I visited Maer and was to get ready for the beginning of September. My uncle (Josiah Wedgwood) sent me to Shrewsbury and spoke to my father in the course of the firing, since my uncle believed it would be sensible to accept the offer from me. I considered it wise.

My father was convinced that he was among the most sensible men around and immediately accepted in the most amiable manner. "that I ought to be clever enough to pay more for my food allowance aboard the "Beagle";" I was required to lavish myself and also to comfort my father He said "But they say that you're very intelligent." He replied with a smile.

The the next day, I was able to go to Henslow at

Cambridge and then to visit Fitz-Roy in London Then all was set. After I had become extremely close to Fitz-Roy I realized that because of my nose's shape I stood the best possibility of being rejected! He was a committed Lavater fan and was convinced that by the outline of his face, he could discern the personality of the person. However, he was wasn't sure if those with my face have the energy and drive to complete the journey. However, later on I think that he was extremely pleased with the deceitfulness that my eyes revealed.

The character of FitzRoy is unique, with many wonderful traits His job was compassionate flawed, audacious,

determined and undeniably strong as well as his partner was passionate and intense. He would go to any kind of hardship to help people who believed that he needed help. He was a wonderful person who had a truly friendly manner of conduct, which was remarkably similar to an honest man, akin to his maternal uncle, which I learned through the minister in Rio, the famous Lord Castlereagh. He must have taken a lot from Charles II since the Dr. Wallich sent me a photo album he'd created and was as similar to the Fitz-Roy that I was. I saw Ch when I was looking through the tag. E.

Stuart Sobieski, the Count of Albania and child of this King.

The Fitz-Roy temper was unfortunate. The morning was generally worst because he could usually observe a tampering in the eye of his Adler. The reason for his actions was clear. He was very nice to me, however it was clear that he had extremely tough conditions of work within the same cabin in particular, following our mess. To begin, in his beginning years of Bahia in Brazil the man defended and praised the conditions of slavery that I considered disgusted by and even told me that an excellent slave master had just visited him. And if they really wanted to live, they could not be a master of slaves. We had several quarrels. I then asked him maybe in a sneering manner to see if the slaves were able to answer in front of their master was worth it? He was angry and unjustly and claimed that since I doubt his words I could not remain together. However, once the incident had spread it was easy to believe that I was compelled to be removed from the vessels in order for the captain to ask him to calm his anger by insulting me.

I was incredibly honored to be accepted to play with them by the entire weapons officers. But Fitz-Roy showed his usual kindness after a few hours and gave me an officer's apology and a plea for me to remain with his.

His character was among the most noblest people I've seen in many different ways.

It was the "Beagle" trip was the most important event of my life. It determined my entire career however, my uncle was willing to drive me 30 miles to Shrewsbury in the manner that only a handful of uncles would do, and also to have a very poor nose. It's been my conviction that I owe this way I think to be the first genuine education. I was introduced to many areas of natural history and my ability to think analytically was therefore boosted although they had been firmly established.

The geological study of the locations explored was much more important because rationality comes into the equation in this case. In the beginning, a new region seems, there's not anything more depressing than the muddle of rocks. However, when you record the stratification and the composition of rocks and fossils at specific points, and still looking for what else is there The light begins to shine, and the arrangement of the entire area is much more understandable.I carried with me the very first volume of Lyell's "Geology Principles', which I carefully read and found that it was the best book I've ever read. The first location I looked at, St. Jago in the island located in Cape de Verde, displayed the obviousness of Lyell's theories of superiority in geology in contrast with some other writer whose work I read or in the past.

My other responsibilities were to collect animals from all kinds of groups, and to determine and break down a portion of the sea, however it's been a waste of time due to the fact that I'm unable draw the vast majority of MS that I created during my travels and also because I lack adequate anatomical details. This has meant that I have lost a significant amount of time. However, when I first began writing the writing of a monograph about Cirripedia within the next few years, I gained some information on the Crustaceans.

I kept a journal for each part of the day. I put in a lot of effort to describe everything I had observed carefully, in a simple manner, which was a good way to go about it. My Journal also acted partially as letters to my family and parts were also sent to England in the event of a mishap.

In contrast, to the active industries and a focused focus on the activities I was involved in these separate studies discussed above lost their meaning. Everything I thought about or read was intended in order to contribute specifically to what I observed or could have seen; and this attitude persisted throughout the course of five years on the journey. I'm certain that it was this trip that allowed me to pursue the things I did in science.

Chapter 5: RESIDENCE DOWN From the 14th of Sept. 1842, until the present TIME 1876.

We found and bought the house following a series of unsuccessful search in Surrey and other areas. I was pleasantly surprised by the lack of plants in this sluggish region, and as such, unlike in the Midland areas, I felt more relaxed in the tranquil and rustic setting. But, as the writer of an German journal the same thing, it's not so secluded as saying that my home is accessible by mule tracks! Our house here has reacted at this point in a manner we didn't intend and that's to be extremely suitable for our kids' frequent visit.

Few people have had the chance to live an ideal retirement lifestyle like ours. We never went any place other than to relatives' homes and sometimes to the beach or somewhere else. We were part of a small group during the beginning of our time in the house and made some new friends, however, I was affected by the excitement of shivering, nauseating attacks. I've been forced to stop attending any dinner gatherings for a number of years. It is a amount of a loss for me as these events continue to inspire me to be in a good mood. I have invited only a handful of acquaintances from science in this area due to the same reasons.

My biggest pleasure and my sole occupation has been scientific studying throughout my life and the excitement of this work helps me forget the pain of my daily life or makes me forget everything else. Therefore, for the remainder of my life, I've got nothing to share aside from the handful of books I've that I've written. Perhaps it's worth sharing a few details of how this happened.

MY MANY Publications.

My notes regarding the volcanic islands discovered during the "Beagle" excursion was published during the first period of 1844. In 1845, I made several attempts to rectify a different version of my "Research Journal that was originally written by Fitz-Roy around 1839. I've always had better sales success, as my first literary child, more than other novels. The book continues to be sold in England and in the United States to this day and has been translated into German, French and other languages twice. This is impressive, a few many years after the first printing because of the huge success of a novel that travels, and especially an science.In England, ten thousand copies of the second edition were sold. My book 'South American Geological Discoveries was published in 1846. In a small periodical that I have preserved, I state that the three geological publications - including 'Coral reefs' have been in production continuously over the last 4 1/2 years "and this is now 10 years since I returned to England. What time do I have lost due to sickness?" How much time have been unable to use due to sickness? 2nd edition.1876. Coral Reefs (Geological Observations), 3rd edition. 1874.)

I began working in 'Cirripedia' around October 1846. While I was on Chile's coast,

The shape I found was a bit odd and was so deep into the Concholepas covers that I had to construct an entirely new sub-order to acquire the shape on its own. In the last few years, along the shores of Portugal and an ally

Burrowing genus was found. I was required to research and decode a variety of common methods to comprehend the anatomy of my brand new Cirripede which eventually led me to the entire community.I continuously researched the subject over the next eight years, and finally published two volumes of a large size on all species that are recognized in living (published in Ray Company. Ray Company.) as well as two thin volumes of the extinct. I am sure the fact that Sir E. Lytton Bulwer thought of me when he wrote Professor Long within one of his books who wrote two books on limpets.

When I worked for this position for eight years, I'm aware of the condition in my journal almost two years after. In this regard I visited Malveryna for a short time in 1848, to treat hydropathically which helped me do a lot of good so that I could return to work when I returned home. I was so ill that I could not attend his funeral or act as one of the executioners for my dear father passed away on November 13 1848..

My research on Cirripedia I think it is extremely interesting because it is not just that I've discovered several new and fascinating forms, but also added and parasitic to males of some kindI've learned about the cement process, even though I was horribly negligent about my cement glasses. This discovery was later completely confirmed. However, at one time my fertile mind was satiated by an German writer.The Cirripedes form a extremely complex and complicated collection of species categorized by classification. The study I did was extremely useful since I was required to learn about the fundamentals of natural classification as described in the "Origin of species." But, I'm not sure whether it Is worth the time consumption.

I was in the process of in September 1854 putting together my huge notepad researching and testing regarding species transformation. I was extremely fascinated at the finding of huge fossils of animals, which were covered in armor, in the present armadillos of the Pampea formation; and, secondly, the method by how closely alied animals travel south across this continent and, thirdly, the ways that closely allies can be replaced by one another. This is due to their South American character of most Galapagos productions, and even more important because they're subtly different in each island group and none of them is considered to be very ancient in the context of geology.

It was obvious that those as well as other aspects could only be explained by the notion that ecosystems were undergoing a gradual change as well as the issue lingers with me. It was also evident however that either either the natural environment, or creatures (particularly plants) can't explain the many instances when the organisms of all kinds were adjusted in a perfect way to their living habits--e.g. a woodpecker, or a tree-frog that climbs trees. or a crook or feather seed to disperse. I am still amazed by these changes and it is almost pointless to make arguments based on indirect evidence that animals had been altered prior to their ability to be explained in a clear way.

After the time I returned to English it came to my mind that the case in Lyell in Geology could get a new light from the whole topic by capturing all the details related to the wide range of animals and plants that were part of nature and in domestication. In July 1837, my journal was first opened. I was operating on Baconian principles and gathered data specifically on domestic

production and printed inquiries, as well as discussions with gardeners and breeders who are professionals and thorough readings on a wholesale level with no theory.. As I review the complete list of literature I've read put together the results, I am astonished at the work I do, with my entire library of books and transactions. I quickly realized that selection is the key to human success in the evolution of useful animal and plant breeds. However, for a long time I've been unable to figure out what species to select that exist in nature. I enjoyed reading "Malthus on Population" in the month of October 1838, 15 months after my study began, and I was prepared to comprehend the struggle for life everywhere in the long-lasting study of the ecosystems of livestock as well as plants. But I quickly realized that it is possible for beneficial distinctions to be sustained in this environment. as a result, new species will emerge. At that point, I came up with a hypothesis with, but because I was so determined not to cause harm, I was hesitant to decide if I'd even write the shortest sketch for a period of duration. I was pleased of writing a brief 35-page explanation, in the month of June 1842, of my hypothesis and then it was extended into more than 230 pages I written very easily and remain have in that summer 1844.

At the time, I was unaware of a major issue which was interesting to me what I might have overlooked it and its solution, excluding that Columbus theory and the egg. The problem is that organic creatures appear to differ in their character when they change from the same genetic stock. The fact that they differ significantly in the manner that species of all kinds are classified into genera is obvious GENERE in families, sub-command family, etc. and I can remember the exact moment on the road that I was driving when I got the answer to my delight It's been quite a while since I've been Down. The reason, I think is that across a variety of sectors of the natural economy, the offspring of the dominant and growing varieties are adjusted.

Lyell suggested that I write my thoughts in depth around 1856. I began writing at a rate at least three times as big as the 'Roots of Organisms'. However, it's just a brief description of the sources I had gathered which I've used to complete roughly half of the work on this scale. However, my plans were thwarted by an individual named Mr. Wallace, who was at the moment in the Malay archipelago in 1858's summer and sent me an essay entitled "On the tendency of Varieties to stray forever from the Original Type;" Mr. Wallace asked me to send the essay to Lyell for a review in the event that I am impressed by the essay.

In the journal 'Journal of the Proceedings of the Linsean Society 1858, at. 45, are the conditions that I agreed to were Lyell and Hooker's suggestions that the abstract of my MS was to be published with a letter addressed to Asa Grey of the 5th September 1857. It was also published with Wallace's essay. I was initially cautious about accepting the proposal since I thought it would be unjust to Wallace. Wallace, because then I didn't realize the kind and noble nature of his intention was. The article in my magazine as well as Asa Gray's letter hadn't been written, and they've been badly written. However the writer, Mr. Wallace's essay was clearly and beautifully written. However, our productions together were not arousing much attention and also Prof. Haughton of Dublin who concluded that everything they learned was untrue and that all that was actually true was old news and outdated, had only seen one that I remember. This illustrates how important it is to communicate every new idea to the general public on a regular basis. In the month of September, 1858, I began researching Lyell and Hooker's firm recommendation to create an account of the transformation of animals, but was interrupted by the declining well-being of animals. I wrote a summary of the MS that began in 1856 and grew to a greater extent and then finished the book at the same size. The process took 13 months with 10 days intense work. The book was released in November 1859, under the title "Root of the Species.' Although it was greatly expanded in subsequent editions and corrected the text, it was the identical book.

It's my main job. It was very good right from the start. At the time of printing it was a limited edition of 1250 copies was sold and following another edition of 3,000 copies were sold. Today, 16,000 copies were purchased throughout England (1876). This is an impressive number considering how expensive a publication it's. The majority of European language that is spoken, including Spanish, Bohemian, and Polish was transposed.

And Russian. and Russian. Based on Miss Bird, it has also been translated into Japanese and a lot is being studied. (Miss Bird is incorrect, according to professor Mitsukuri.--F.D.) There was an article written in Hebrew that proves that the concept is found located in The Old Testament! There were many reviews. I kept track of everything that appeared in "Roots" and in my other relevant books, and the total (excluding review articles in journals) to 265; however I gave up the project in the end of the day in desperate need. There were numerous articles and books on the subject. A Bibliography or catalog of "Darwinismus" was published in Germany every year or so.

I believe that the acclaim of the Origin might be due to my writing two sketchy sketches a while in the past and then abstracting a larger manuscript that was abstract. This allows me to select the truths and conclusions that are the most striking. I've been following the golden rule for many years. That is, whenever a truth that was publicly announced came to me as a fresh realization or reflection that was contrary to what I had observed in general I've learned from experiences that these particular facts and emotions are more likely than the ones I like to disappear from my memory. To write down a note of the information without fail and quickly; Due to my habits, I'd been unable to hear and attempted to address only a few questions about my views.

There was a common belief that success in "the source' proved that the origin was out of the blue, and that the minds of men were prepared for this.' This is not necessarily true but I'm not sure if I believe it because I did not always appear to sound like some naturalists and have never met any who appeared to doubt the existence of the species. The two of them, Lyell and Hooker were not able to agree with me, even though they were listening to me with curiosity. Then I attempted to convey to intelligent men a couple of times what I meant with Natural Selection. What I believe to be real was that an abundance of observed facts were ingrained by naturalists who were ready to accept their place until clarification was provided to a idea that would lead them to the conclusion. Another reason for the book's success was its size due to the late Mr. Wallace's essay. If I'd written it on the book that I was beginning to write at the time of 1856, my work would have been about four or fifty times as large as the "Origin," and only a few people would have been able to keep the process of reading the essay.

In publishing around 1839, at the time that the concept was evidently created around 1859. I was able to benefit greatly; however, I did not sacrifice much because I was concerned at minimally if those who wrote it would assign myself or Wallace the most originality and his work certainly helped in the development of the idea. I've been entangled to one specific aspect that has always irked my pride. That is the clarification during the Glacial period of the presence on mountain summits that were remote and in the arctic regions of the same species of plants as well as a few animals.I am so glad to have written it down and, I believe Hooker took it up several years prior to E. I'm thinking. His popular memoir ("Geologue. Survey Mem.," 1846) was written by Forbes. I am still convinced that I was right on very few issues on which we didn't agree. Naturally, I did not mention in the paper that I had discussed on this point in a separate manner.

There was not much joy when I wrote 'Origin when I was writing about the huge gap between embryos and adult animal in some classes, and the close relationship between the embryos at similar age. In the initial reviews of the book's 'Origin', a note was included to be true, as I remember and I recall writing an open letter to Asa Grey regarding this point. In the last few the last decade, Fritz Muller and Hackel were recognized by numerous critics, who would have certainly been more involved in the process.

thorough, and more precisely in some ways more precisely than I could have done it more thoroughly and with greater accuracy than. I had all the material for an entire chapter on thissubject, and the subject could have been more extensive since I didn't surprise my readers. It is evident that to me, any person who was able to do so deserves credit for it.

I can conclude that most of my readers have been fair to me and have brushed aside other reviewers who do not merit of their attention without having a solid understanding of science. My view was often disseminated, criticized and ridiculed, however the majority of them did it on good terms, which is what I am of the opinion. In all I'm convinced I've consistently overvalued my work. I am happy that I avoided controversy. I must thank Lyell who, a few years ago, had firmly advised me not to be involved in an argument over my geological research, since it never had any impact and was a terrible wasted of time and attitude.

I recall when I was I worked in the Successful Performance Bay, in Tie I can remember the time the times I felt weighed down or was working in a way that was unsatisfactory and I was ridiculed with scorn as well as when my performance was judged to be overrated and was then ashamed, and it was the worst thing I've ever experienced.

It is a pleasure to tell me hundreds of times, "I have worked as long and as hard as I can, and nobody can do better than I have done." I've done it in the fullest capacity and even opponents can claim whatever they want, but my conviction is unbreakable. For the last 2 months in 1859 I was incredibly involved in the process of planning and executing.

correspondence from a second edition of the book 'Roots.' One January in 1868, I began to organize my notebooks for book regarding The Variety of Domesticated Animals and Plants and it was not published until the beginning of 1868. It was due to numerous illnesses, including one which lasted for seven months and in part, by the desire to write about different subjects that I was at the when I was most interested.

My little book about fertilizing orchids, which took me ten months of work and was released 15 May 1862. The majority of the data had been taken in over the years. The summer of 1839 was a time of great interest I believe that I was led by the previous summer's experience to cross-fertilization in the flowers by insects, I concluded that crossing had a major contribution to the retention of these methods in the theories on the source of the plants. Every summer after I stayed up to date on this subject in a way and my fascination with it increased dramatically through my reading and acquisition in November 1841.

At Robert Brown's suggestion A C.K. copy. The book of the great author, The Geheimniss der Natur (Das entdeckte Geheimniss der Natur) was the ideal way to write an article on the plant community of many years prior to 1862, rather than relying on the enormous amount of material I'd slowly collected with respect to different species of plants.

My choice was a sound one, as a myriad of intriguing papers and separate research has been published after the release of my book about fertilizing various flowers and that's a lot greater than what I would have thought likely. Since his death the benefits of old Sprengel have been thoroughly examined.

recognized.

I was published in the "Linnean Society Journal' the same year published the 'Two Kinds"On Primula," and five other articles on trimorphic and dimorphic plants in the coming five years. I have never experienced so much pleasure in my scientific life than to understand the nature that these species. I think. I came across Linum flavum dimorphism between 1838 or 1839, and decided that it was a reprehensible variation.

While looking at in the Primula I common species, both forms were more common and permanent than could be observed. I was convinced that the common primrose and cowslip were both in dioetia's path which was with the same shape, the pistil's small size was

In the other, short stamens, they were able to induce abortion. Therefore the plants were put to test in this area but abortion was a possibility that hit its head, until flowering was being fed by short pistils, which were fertilized by pollen of short stamens and produced more grains than any of the four possible unions. Further

The experiment revealed that both kinds have the same relationship with each in the same way as the two genders of a normal animal, since they're both perfect hermaphrodites. There is a more amazing examples of three shapes that are in close contact with each other and Lythrum.I discovered that a similar and unique contrast with hybrids from the union of two animals that belonging to the same type was created by an association between two plant species. I wrote a long essay in the field of "Climbing plants" in the fall of 1864, and then sent this to Linnean Society. It took me four monthsto complete, however, I was so numb when I learned the details that I was forced to leave the work in very bad shape and at times in a dark way. The work was not well-known however, when it was corrected by 1875 and released, the paper was released in a separate document. I was influenced by Asa Grey's brief paper published in 1858 to discuss the subject. He gave me seeds and I was amazed and confused by the stems and stirrings that are actually quite simple steps but appear complex at first glance and I was able to identify types of climbing plants. I had some plants. Then, the whole subject was researched. I was drawn to this because I was completely unhappy with Henslow clarified his understanding of two plants during his lecture specifically, they appeared to ascend into an edifice. This was not the correct explanation. All of these modifications are as impressive as Orchids to ensure cross-fertilization climbing plants.

As previously reported at the beginning of 1860 my book 'Variation of Animals as well as Plants under Domestication It was not finished until 1868, the year I began writing it. It was an amazing book that required an effort for four years, and 2 months. It contains all my opinions on our domestic products as well as a vast amount of information gathered from a variety of sources. As far as current knowledge allows, the reasons and the reasons for variations, heritage as well as other issues. are examined in the 2nd Volume. I will present my used Pangenesis theory at the conclusion of the book. The unsubstantiated theory is of very little or no value, However, if any one was to later make remarks which could lead to the creation of such theories that I have offered, I will have provided a solid argument because an enormous amount of evidence that is isolated could therefore be combined and made more understandable. Another version was published in 1875, which was mostly modified and I found it expensive.

In February 1871 , my "Descent of Man" was composed. I could not resist the idea that human beings had to be controlled by the same rules until I was convinced by 1837 and 1838 organisms are mutable. I then collected information on this subject for my own pleasure and didn't publish them for a considerable period of time. Although the derivative of a specie is never mentioned as the 'root of the species' discussion I thought it was appropriate, for anyone to not be able to accuse me for obscuring what I was thinking, to state that "light will be thrown at the origins of man and his past." It was unjust and also unjust.

negative to the popularity of the book because it reveals my beliefs in relation to its source without providing details.

When I found out that a lot of naturalists had believed in the theory of evolutionary speciation I decided to make notes on my own and create a distinct document on the origins of humans. I was elated to write this as it offered me an opportunity to investigate sexual selection completely -- an issue that always intrigued me. The only subject I could write about is variations in our goods and the reasons and regulations of differences in the plant world and cross-linking of heritage using all the sources I've gathered.The "Descent of Man" required me to write for three years in order to write, and often it was a case of ill health was lost, and other were absorbed when new editions and smaller works were designed. The book was released in 1874, in a revised and more reformed edition.

The book was composed in the fall of 1872, and was the basis for my book "Speech of emotions" was published. in animals and humans.' A single chapter on the subject was planned to me within the "Descent of Man" but when I began to put all my thoughts together I realized the need for a separate book needed.

Chapter 6: I just loved shooting! But I believe I have to

I was not aware of my enthusiasm because I was trying to convince my self that it was an analysis-oriented job. It required a lot of expertise to determine where the majority of the game might be located and to follow the dogs well.

The memory of one of my fall trips in Maer in 1827 was vivid, just like the story I'd read by Sir J. Mackintosh. I was enthralled with a sense of happiness when he stated that he had a certain impression I've observed about the young person.' It's probably because I was able to hear every word he spoke with great interest as I was unsure what his subjects, such as political, history and moral philosophy of the time were all about. It's very good for a young person to be able to receive praise from an influential person although he might likely or certainly excite the ego, as it helps to keep him moving in the proper direction.

In the next 2 or 3 years,, my excursions to Maer were extremely enjoyable regardless of the autumnal shooting. The countryside was a blast to explore or ride around as well as in the evening it was a great place to have enjoyable conversations, but not as intimate, like there are large family celebrations with live music. The pace of life was relaxed there. The entire family would be seated at the front of the portico that was built in the past with a garden with flowers as well as a high wooden bench that ran across the bay with fish thriving in the midst of the bay and a water-bird paddling. Within two or three years when I visited Maer the shooting in autumn was very enjoyable. There was plenty of enjoyment cycling or walking through the countryside. In the evening, there was plenty of great conversations, but not as intimate as typically big family gatherings and music. It was a completely unstructured and relaxed atmosphere. In summer, all the family would gather around the portico's old structure, which had the rose garden and a soaring wooded bench overlooking the bay. There were sea birds as well as fish soaring up and down.

In retrospect, I see the consistent dominance of my love for science over other interests. My love of shooting for a long time has been able to last for almost all of my initial two years of shooting and I shot on every bird and other animals that I could find However, I had to hand my weapon over repeatedly, and eventually my servant as shooting harmed my work, and, in particular, the structure of geology in an area. Even though I was inconscient and ignorant, I discovered that the pleasure of observation and thinking is far more satisfying than the enjoyment of being competent and sports. The fact that my mind was shaped by my travels was evident from a statement that my father made an extremely keen observer I was seeing and who had a cynical stance that was far from the perspective of the phrenologist. For after seeing me, he looked over to my sisters saying, "Why, the shape of his head is different." When he saw me at the end of my trip flipped around.

The ride is back. On September 11, (1831) I went to"the "Beagle" located in Plymouth together with the Fitz-Roy. As I travel to Shrewsbury I say a heartfelt goodbye to my mother and father. I moved into my house in Plymouth on the 24th of October. continued to stay there until the "Beagle went off for the rest of the world to sail the beaches of England. We had tried to sail twice prior to that, but were supported by the wings of a strong bird each time. The two months in Plymouth as I was unable to complete my studies in all aspects I practiced.I decided to leave all my friends and family for too long. The weather was just too foolish for me. I was also afflicted by the pain and palpation in my heart as a lot of young men in the dark were convinced that I had a heart condition and especially someone with only a few bits of medical expertise. In the midst of waiting to hear the verdict that I was not ready for my trip, and I determined to go regardless of the odds I didn't visit any doctor.

I'm not required to write down the travel experiences such as where the trip was and how we spent itsince in my written Journal I've provided an extensive account. The beauty of the tropical blooms are more vivid than all other things that I can recall in the moment; as I hold a resounding memory of the sense of sheer awe which the magnificent Deserts in Patagonia as well as the forest of Tierra del Fuego mountains aroused in me. A view of a naked wild animal in his home country is an unforgettable experience. A lot of my horseback trips in the open or on warships in which some were for weeks in length, were

It was fascinating to note that It is fascinating that they weren't a problem in the present or later. I am still thinking of the issue of coral islands and the geological makeup of these arelands, for instance St. Helena, with

A great deal of pleasure from certain aspects of my research. There is no need to go into the details for all people in South America of the special connections between species of animals and the plants that inhabit the numerous islands in the Galapagos archipelago.

So far as I'm able to assess myself, I worked to the max during my journey from pure enjoyment of inquiry to the huge amount of information in the natural sciences. However, I was confident that scientists could achieve an adequate position, whether they were ambitious or not. Unlike my colleagues I am not able to give an conclusion.

It's a striking and simple Geology of St. Jago: A stream of lava was once baked into the form of a hard white stone, comprised of crushed shells as well as corals. It was all lifted at the time. But the white line of rock provided me with a fresh and important fact that the craters that were in active use were subsided and there was lava pouring out. This was the moment that I could create a book about the geology of the various nations I had visited. It was with great joy that I was thrilled. It was a great experience.

An unforgettable experience for me. I remember vividly the low point in which I lay, while the sunlight was brightly and some strange desert trees soaring above me, and the live corals that sat at my feet in the tidal waters. In the course of my travels I was contacted by Fitz-Roy to review a portion from my Journal and declared that it was be worthy of being published. After our trip , I received an email from Ascension which informed me that Sedgwick called my father and stated that I was one of the top scientists. I was unable to imagine what Sedgwick could have learned about my research and I was told (I remember in the future) that Henslow was reading a few letters I wrote to him prior to his visit to Cambridge

Philosophical Society (Read at the meeting that was held on November 16 1835 and published in a pamphlet comprising 31 pages to distribute to participants of the Society.) The pamphlets were made available for private distribution. The fossil bone collection I I gave to Henslow was also of palaeontologists lot of attention. I climbed with a bordering step over the Ascension mountains, and played with my geological hammers to rock the volcanic rocks. It's clear how ambitious I was However, I suppose I'm not able to claim that over the years, when I've looked to be able to applaud men like Lyell and Hooker to the greatest extent possible. I didn't have any concern for the general population, for whom I was my acquaintances. I'm not saying that I didn't appreciate the glowing reviews or a large sales volume of my work too much but it was a momentary pleasure and I'm certain I've never moved one inch off my path to fame.

From my return to ENGLAND (OCTOBER 2 1836) to my wedding (JANUARY 29th, 1839.)

Three and two months were among the most busy year I've ever had and I was in a state of discomfort and spent a lot of time. The 13th of December was when I stayed in a hotel

Cambridge (in Fitzwilliam Street) after I walked back and forth between

Shrewbury, Maer, Cambridge and London in which my entire collection was under the direction of Henslow. In the past three months, I was living here and Professor Miller helped me analyze my rocks and minerals.

I began to write My Travel Journal, which was easy because I had a good MS Journal. MS Journal was written carefully and my study produced an abstract of my most important scientific findings. I've sent my research findings about the height of Chile's coastline to the Geological Society shortly at Lyell's invitation. (The Proc. Geological Proceedings, Soc. Proc. ii. 1838, 446-449.)

I was there for nearly two years prior to getting married on the 7th March 1837 in Great Marlborough Street in London. In the past two years, I've been writing my Journal and reading articles prior to joining I joined the Geologic Society, writing the MS of my observations in geology and preparing on the release of "Beagle" The month of July, I started my first notebook about things that pertain to the evolution of living organisms. I have concentrated on for a lengthy time , and never stopped to think about it throughout the following twenty years.

I was also a member of the society for two years I was one of the honorary Geological Society Secretaries. Lyell I've met him many times. His compassion for the efforts of others was one of his most prominent traits and I was just as surprised as I was encouraged by his genuine concern upon my return to England. My thoughts about coral reefs was told him about coral reefs. This greatly inspired me and his guidance and example influenced me deeply. At the time, I saw some Robert Brown a lot; I would always call and sit at his table on a Sunday mornings during breakfast. He spilled out an extensive collection of insightful thoughts and sharp observations however, they almost always linked to a series of questions. Over the course of those two years I went on a few trips for rest, as well as a more extensive excursion to the parallel roads in Glen Roy, whose account was written within"Philosofical Transactions "Philosofical Transactions" (1839 Pages 39-82). Because I was deeply

Incredulous at what I had discovered fascinated by what I had seen about South America, I attributed the actions of the sea due to parallel lines. However, when Agassiz presented his theory regarding the idea of the glacier pool I was forced to reconsider this theory. In the time when we were in a state of understanding , no other explanation could be possible, I made arguments for marine activity; and it was my error to put my faith in science and the concept of exclusion.

Because I could not perform science every day In the past two years, I read quite a lot on various subjects and texts, including several philosophical ones. At this point, I was enthralled by poetry by

Wordsworth as well as Coleridge and I can boast of having read the 'Excursion' two times. The 'Paradise lost' of previously Milton was my most favored and I've always preferred Milton on my visits to the Beagle in the days when I was able to only carry only one volume.

My wedding day was january 29th, 1839. RESERVATION IN UPPER GOWER STREET to our departure from LONDON and settling at DOWN on the 14th of September 1842.

I was not as involved in research in the three years and eight months that we were in London however, I worked the best I could just like I did in every other similar period in my lifetime. The reason for this is frequent malaise and a lengthy and long-lasting illness. My research with "Coral Reefs "Coral reefs" that I began before my marriage and that was revised in the final proof sheet in the 6th of May 1842, was focused on for a lot of my time, even though I had no time to do anything. As a young man I read this book, it took me 20 months of labor as I had to read each article about the Pacific Islands and consult a large number of maps. Scientists were very involved The idea is well-known I'm sure.

None of my other works was conceived in this way, since the entire concept was first formulated in the western part in South America before I saw the coral reef in its real form. Then, I needed to conduct a thorough investigation of living reefs in order to verify and extend my theories. But, it must be noted that I regularly observed over the past 2 years about the irregular rising of the land, and of sediment depletion and denudation along the coast from South America. I was forced to think about the effects of subsidence, and the ongoing accumulation of sediments caused by the growth of corals upwards was quickly replaced in the imagination. This was the basis for my theory of the barrier atolls and arcs.

Alongside my research on coral reefs for my dissertation I've also read the Geological Society records on the Erratic Boulders of South America ('Geology. Soc. Proc. iii. 1841) during my stay in London during my time in London, I experienced the earthquakes ('Geology. trans. V. 1840) as well as the genesis for earthquakes throughout Mold from the departments. I've been able to supervise the publication "Zoology of the "Beagle Flight.' ('Geology. Soc. Proc. ii. 1838)

I have not disrupted the collection of data about the species' ancestry however, I may be able to do this due to the fact that I can't be a part of something else than disease.

The summer of 1842 was to research the effects of the glaciers that once dotted the larger valleys. I had been more healthy than I had been for a long time, and traveled through northern Wales. This trip was truly intriguing to me, and was the only time I felt physically fit enough to climb a mountains or trek for long distances , as is required for geoscientific research. Then I wrote down a short account of the experience within The Philosophical Journal in 1842.

I was determined enough to be a member of an all-inclusive society at the start of my time in London I was able to meet many men and scientists who were more or less distinguished. A few of them, even though I don't have anything worth mentioning, I'll offer my

observations.

I've met more Lyell in the years before and after my wedding as compared to any man. He was an incredibly handsome man.

easy, thoughtful approach, good judgment and a lot of originality, it appeared to me. When I wrote about Geology with him

He didn't stop until he had clearly understood the entire picture, always giving me a better explanation than I had. He would not agree with my idea and, even when he was exhausted He would be doubtful for a considerable period. His genuine empathy for scientists of other fields was another aspect. (The note on Lyell were added on April 18, 1881. This was just a couple of years after the majority Recollections published, in order to explain the small amount of duplicates that can be seen in this article.)

After returning from my "Beagle" travels my opinions regarding coral reefs, which differed from his were explained and the fervent interest that he showed was awe-inspiring and impressed me. He was passionate about science and believed that the advancement of civilization was of most significant thing to happen. With regard to his theological beliefs or in his disbeliefs He was incredibly kind and utterly liberal. He was a well-known theist, but he was also a very honest one. He was quite remarkable in his honesty. He proved that he was a shameful individual, however by disagreeing with the views of Lamarck, he gained a lot of fame. He became older. "What is a great idea for every scientist to die at the age of 60 years old, since afterward he'd be certain to be against all new ideas." He reminded me a while back when he mentioned the geologists of the past and their opposition to his current beliefs.

Geology research is owed by Lyell an immense debt - far more likely than any other human being who has ever lived. When I embarked to explore the 'Beagle', the charming Henslow who was at the time convinced that he was a geologist, just like every other that there were cataclysms after another I was to purchase an early volume of the "Principles' which was not yet composed. What a difference will people be able to learn about the 'principles' of today! The first place I'm proud to have that is St Jago from the Archipelago of Cape de Verde that I geologized, convinced me of the supremacy of Lyell over any other job I was taught.

Lyell's influence is evident in the many technological advancements in France as well as England. Lyell is primarily due to the complete disregard of the wild theory about Elie De Beaumont including "Elevation Craters" and "Elevation ranges' (the latter concept I saw on Sedgwick in The Society of Geology, lauding the sky).

I've met Robert Brown a fair deal and he was known as Humboldt, "facile Princeps Botanicorum," He appeared to me to be particularly impressive due to the exactness and exactness of his discoveries. His knowledge was extremely high , and he passed away often because of the fear that he would have a misstep. He shared his knowledge in a surprising manner however, he also was a little at times jealous. He demanded that I examine an eye-piece and, on occasion to explain what I observed. I had spoken to him a couple of times prior to the "Beagle" trip. This I did and I now believe that in a vegetal cell there was amazing protoplasmic currents. Then I inquired about what I had seen but he said, "That is my little secret." I was told by him.

He'll do the most charitable acts. When he was sick, extremely unhealthy , and not suitable to undertake any project (as Hooker told me) was a regular visitor to an elderly servant who was a long distance away (and whom he helped) and read to him aloud. This was enough to compensate any difficulties in science or jealousy.

I'll try to list some other notable figures I've seen occasionally however I do not have anything to add. Sir J. Herschel was highly respected to me, and I loved dining in his lovely home in Good Hope Cape and later in his London home. A couple more days, I saw him, too. He didn't speak much, but it was interesting to listen to each word he said.

I had the pleasure of meeting Humboldt who was gratified by sharing his expectations of seeing me, during breakfast at the house of Sir R. Murchison. I was dissatisfied with the man but I certainly was not expecting too much. There's nothing I can remember about our conversation, aside from the fact that Humboldt was extremely pleased and spoke quite a bit.

Remembers Buckle whom I had the pleasure of meeting in Wedgwood's Hensleigh once. I've been delighted to learn his system for collecting data from him. He informed me that he had purchased every book that he read, and made each one a comprehensive list of information he considered useful to him. He also stated that for every book that he read that he remembered, he could recall it because his memory was amazingly good. I asked him to first identify advantageous facts. He said he was unsure however he was driven by a certain type of instinct. Based on his long-standing tradition of creating indexes, he was able to incorporate the staggering quantity of sources covering various subjects within his account of civilization. I found this book extremely fascinating and have had it read twice, however, I'm not sure if the book is worth it because of its generalizations. Buckle was a great speaker and I never heard him speak or thought. I couldn't have done that, as there were no gaps. When the lady. Farrer started singing, I leapt up and shouted, "Well, Mr. Darwin's books are more interesting than his conversations I'll definitely be able to hear her when I've died and he'll show up to his companion.

I came across Sydney Smith in Dean Milman's home once, along with other famous literary characters. In every word Smith said there was something incredibly humorous. Perhaps because of the excitement to be amused. Lady Cork who was actually old in the moment was thinking. This is the woman who was profoundly affected by a sermon that she took the guinea of a friend so that she could set her tray on it. He later said, "It is generally believed that my beloved lady friend Lady Cork was not noticed," and he said that in a manner that for a brief moment, no one could dispute that it was because the devil was overlooking his beloved friend.How he was able to convey this, I don't know.

I even had the privilege of meeting Macaulay once at Lord Stanhope's residence and, as there only one other person eating dinner, I got a the chance to hear him talk and was a lot of enjoyable. Macaulay didn't speak too often; he could not speak about the man, so long as he let other people alter the flow, and he let it happen.

Once upon a while, Lord Stanhope provided me with a small proof of Macaulay's accuracy and completeness: a lot of historians would gather at the house of Lord Stanhope and would discuss a variety of subjects, they frequently distinguished them from Macaulay and, in the past they'd refer to a different work to see who was correct.

In another instance I had the pleasure of meeting one of his historians and literary gatherings, Motley and Grote, at

Stanhope's home. After lunch, I spent about one hour Grote strolling through Chevening Park and he was enthusiastic about the chat, and happy that his manner of speaking was simple and unpretentious.

In the past, I enjoyed dinner with the old Earl of the father of historians. He was a strange man however I loved of what I learned about him. He was honest charming, brilliant and sweet. His characteristics were easily identified with a brownish hue, and his attire was completely brown when I met him. He seemed to be a believer in something that seemed absolutely unbelievable to some. One day , he told me "Why should you just give up on your faff-faddle with the sciences of zoology and geology and concentrate on the mysterious sciences!" Historiographer Lord Mahon turned his attention to me and his lovely wife was extremely entertained.

The last person I'd be able to remember was Carlyle I saw him at the home of my brother numerous times, as well as at my home twice or three times. The conversation, as well as his writings, was intriguing and racial however, he was too deep about the same subject frequently. My brother was funny during the meal and Babbage and Lyell who all loved to talk about and debated with others, were among them. But, Carlyle silenced everyone by arguing about the advantages of secrecy throughout the dinner. Carlyle was thankful for the lesson he had learned on the quiet after supper. Babbage in his most savage appearance.

Carlyle was a snowman for almost everybody in my household. One day at home, Carlyle called Grote's "Culture" as "a sexually ferocious quagmire that has no spiritual resemblance to it." I have often believed that his sneers could be interpreted as laughter before the 'Reminiscence' came out but that seemed doubtful. The man was described as a sad and sad, yet generous man and his heart that he laughed with is well-known. I am convinced that his goodwill was authentic, and not even a hint of envy stained his character. It is not difficult to see his incredible ability to visualize things or people more than any Macaulay image, it appears to me. It's a different matter if his photographs of men are genuine.

He was all-powerful and could impress men's minds.

These are the great spiritual truths. His opinions on slavery were fierce but on the other hand. Perhaps it was in his head. His thinking seems very limited to me, despite the fact that the field of study he dismissed. It's wonderful to have been able to describe Kingsley as a seasoned scientist. He was chuckling at the notion that Goethe's ideas about light could, as I stated be evaluated by mathematicians like Whewell. He believed it was the most absurd thing to consider whether glaciers were moving or speedier or slower. I have never encountered a person who had such a sloppy thinking style in scientific research to the point that I could be able to judge.

In my time in London I was able to attend meetings of various scientific societies as often as I could. I also was secretary for the Geological Society. But the amount of time I spent in presence, and that normal culture was so harmful to my health that I decided to remain in the nation we both wanted and respected.

I began to take notes about the very first emergence of various expressions he showed. In the beginning, I believed that the most complex and exquisite shades of speech came from an unchanging and regular basis. My birth date was the 27th of December, 1839. The season of summer 1840 I was reading Sir C. Bell's fantastic expression works, which significantly stimulated my fascination with the subject, however, I was not able to believe that the different muscles were created for the purpose of

expression.. I have been a bit involved with the subject from that time onwards. The human and the domestic animals are all part of the. My book has sold a good quantity; as of the date of publication the book, there were 5267 copies available.

The summer of 1859 I ran and relaxed close to Hartfield which is home to an plenty of two species of Drosera and also saw many insects trapped from the leaf. I took some plants back and watched the tentacles move when I fed them insects.

This led me to believe it possible that due to an underlying reason, the insects were kept. I was fortunate to discover that lots of leaves were placed in separate fluids that were nitrogenous and not with similar volume. When I discovered the first by itself to trigger rapid motions It became apparent that there was a great new research area here.

I went back to my experiments in the subsequent years during my downtime and wrote my book on insects in July 1875, a full sixteen years following the first time I made observations. The time pause was an enormous benefit to me in this case and in all of my other works; for one can review his work after a prolonged period of time, just like the work of a human. It was certainly an interesting discovery that plants can produce, when it is well-agitated it, an acid-and fermented liquid that was similar to the fluid that a pet's digestive system produces.

I will write about the self-fertilization and the cross-pollination of the Kingdom of Vegetables in this fall of 1876.

.' This book will be a supplement to that volume "Fertilization of orchids" in which I demonstrated that cross-fertilization techniques are effective and I'll show how important the results are. I was led to carry out various experiments within the volume for a period of eleven years without any intention.

observation. In reality the moment my attention was attracted by the unique nature of the fact that the seedlings from the autochthonous parent have lower heights than the autochthonous parent, the incident required a repeat, for the first time, of parentage seedlings that have been cross-fertilized, they are significantly size and the vigor. I'm also planning to publish my revised Orchids book as well as my dimorphic and trimorphic species. I will also add some additional comments on the allies I haven't had time to study. In the meantime, my energy will be depleted and I'll be able to call to "Nunc dimittis."

Written May 1st 1881.

The findings of this study can be explained as I believe the endless and amazing techniques used to transport pollens from one species to the next within the Genus. "The Effects on Crossing and Autonomization This book was published in autumn of 1876. However, I feel I could have emphasized more strongly than I had done about the different ways to self-fertilization, particularly according to the comments by Hermann Muller; while I was well aware of a myriad of modifications. An even larger version was composed on 1877 in my book 'Fertilization of Orchids.'

The year 1880 witnessed the debut of "The diverse forms of flowers, etc." This book is primarily made up of several articles on heterostyled flowering that were originally printed through the Linnean Society, with a number of novel substances being added and a few other discoveries in the case of two varieties of flowers within one plant.

As I've mentioned in the past, I've never enjoyed my little explorations as deeply as the heterosexuality that exists in flowers. I think the implications from the illegal crossing of certain species of flora, as they are tied to the sterility that is associated with hybrids to be very important, however, only a handful of people have a clue about it.

It was my intention to write my version of the "Life of Erasmus Darwin" of Dr. Ernst KRAUSE in 1879. I added a sketch of his characteristics and methods that were followed by Dr. KRAUSE that I have in my possession. There were many people affected by this short story and I am astonished to learn that it was only 800-900 copies that were sold.

In 1880, I published our book 'Strength of Motion in Plants' using the assistance of my sonFrank. Frank. It was a laborious process. The book bears much the same relation to my small book on 'cross-fertilization' and 'orchid fertilisation;' since it was difficult to account for the growth of escalating plants in too many broadly distinct types, according to the philosophy of evolution, except for the fact that each type of plants have a certain light movement ability of the same nature. I have demonstrated this, and, furthermore, I've been guided by a broad generalization.

Apparently. The important and wonderful types of motion, triggered by light, gravity attraction, etc. All of them are modified aspects of the circular movement. It's always fun to praise organized plants, and I was thrilled to demonstrate the variety and appropriate movements a root tip has.

It's of little importance, though I'm not sure if people is interested in this particular issue (Between the months of November and February, 1881: 85,500 is an example of how the amount of sales) It has an interest for me. I've put together a small book on the evolution of vegetable mold through the operation from Worms.' The book was published over forty years ago, when I presented a short paper presented to that of the Geological Society and resurrected old geological ideas.

Since then, I've read every book I've written that were landmarks of my life, and I have no more to say. I'm not aware of the past 30 years of any shift in mind, with the exception of an issue that's currently to be discussed. Nor is any future change anticipated in the event of a global change in the circumstances. However, my father had his mind as clear as ever throughout his 83rd year , and all his talents were the same The only thing I want to do is pray that I attain a certain degree of consciousness before my mind is gone. I believe that by forming the correct explanations and conceiving of tests, I've gained a bit of professionalism however this could be an outcome of my own experience and a higher level of knowledge.

It has led to me with a huge loss of time but had the added benefit of forcing me to think thoughtfully and deeply about every sentence, as my logic and my own beliefs or the opinions of others have caused me to make mistakes. I've not always been able to be the best at it. I haven't always faced the same problems as me.

There is a mental defect that is the cause of me misrepresenting my argument or idea. Before I wrote my sentences, I was concerned about my sentences. Later, for a long time I discovered it was more efficient to use time to write pages in a hurry with my sloppy handwriting, by contracting the words, and then fixing the rest. Therefore, sentences written in scripts are always longer than what I could intentionally write.

After a long discussion about my work I'm now suggesting that in my most popular books, I spend much time thinking about the topic in general. Then, in about two or three pages I sketch out the basic outline. Then, in the space of a few pages, create a few words or a single word for an overview or a series of facts. Each section is expanded and changed frequently until I begin writing in depth. The information that was used in my books from other authors were often utilized and I had several diverse subjects I can say that I have between 30 and 40 large portfolios.

I can place a separate memorandum or reference in cabinets with shelves that are numbered. I've ordered a variety of books, and I am keeping track of all the specifics that impact my work after they have been completed and if the book is not my work I'm writing an abstract that is different and I have a large drawer full of abstracts. Before beginning any topic I go through the short indexes of all books and create an overall and categorical index and am armed with all the information I have accumulated throughout my life, ready to be utilized, putting it into one or two portfolios that are appropriate. In one sense I've stated that in the last 20-30 years, my thinking has changed. When I was thirty or more, I had the immense pleasure in being a poet of all kinds and styles, such as poetry by Milton, Gray, Byron, Wordsworth, Coleridze, and Shelley.Of the classic plays, particularly. I've said before that prior pictures brought me a lot of pleasure as well as songs. Even though I've been unable to stand the thought of reading poetry for a while currently. I attempted to read Shakespeare in the past, but found it to be so boring, I became sick. For pictures, or songs, I've lost my interest. Music is usually a way to concentrate about what I was doing instead of pleasing me. I have a certain appreciation for beautiful landscapes, however I don't get the same joy and satisfaction that it once did. However,

For a long time, the novels that are works of imagination have always been a wonderful relief and pleasure for me, even if they're not the highest quality, and I am always thankful to writers. I have read an incredible amount of books aloud. I like to read it if it's well-written, or is not a happy ending and that is a rule that should be avoided. The book is not brought in the first class in the manner I'd like it to except if it contains somebody you'd love to spend time with and if it is a beautiful woman is ideal, so much the better.

The eerie and tragic absence of higher esthetic preferences is the most bizarre thing of it all: I've got the same amount of curiosity as I've ever owned books on biographies, history and travelogues (regardless of the evidence they provide) as well as essays on a variety of areas. My mind is an electronic device that can breach the law of common sense by accumulating a vast array of information, yet it's impossible to imagine that this might have led to the degeneration of this part of the brain that is based on higher-end taste. If someone had a more organized, higher brain than mine would not have been affected in the same way I'd like to listen to poetry and certain music at least once during the week, if I were to live my life once more as if the weakened parts of my brain could have been kept an active state from the use. The loss of these pleasures can be a loss of pleasure and could affect the brain as well as the spiritual through reducing the emotional component of our lives.

www.ingramcontent.com/pod-product-compliance
Lightning Source LLC
Chambersburg PA
CBHW050401120526
44590CB00015B/1778